Managing

D0705780

Managing and Delivering Performance

By
Bernard Marr

AMSTERDAM • BOSTON • HEIDELBERG • LONDON
NEW YORK • OXFORD • PARIS • SAN DIEGO
SAN FRANCISCO • SINGAPORE • SYDNEY • TOKYO
Butterworth-Heinemann is an imprint of Elsevier

Butterworth-Heinemann is an imprint of Elsevier
Linacre House, Jordan Hill, Oxford OX2 8DP, UK
30 Corporate Drive, Suite 400, Burlington, MA 01803, USA

British Library Cataloguing in Publication Data
A catalogue record for this book is available from the British Library

Library of Congress Cataloguing in Publication Data
A catalogue record for this book is available from the Library of Congress

ISBN: 978-0-7506-8710-2

For information on all Butterworth-Heinemann Publications
visit our website at books.elsevier.com.

Typeset by Charon Tec Ltd., A Macmillan Company. (www.macmillansolutions.com)

Printed and bound in Great Britain

09 10 11 12   10 9 8 7 6 5 4 3 2 1

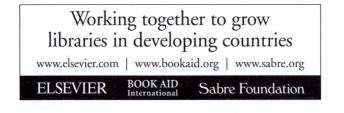

To Claire, Sophia and James, the three pillars of my happiness

Contents

About the Author   ix

Preface   xi

Introduction – Managing and Delivering Performance   1

Part I – Identifying and Agreeing What Matters

1   Clarifying Purpose, Goals and Values   19

2   Understanding your Outcomes, Outputs and Value
    Proposition   29

3   Understanding Inputs, Competencies and Resources   45

4   Mapping and Defining your Strategy   67

5   Aligning your Organization with Your Strategy   107

Part II – Collecting the Right Management Information

6   Measuring Performance   135

7   Creating Key Performance Questions   161

8   Designing Performance Indicators   175

Part III – Learning and Improving Performance

9   Fostering a Performance-Driven Culture   211

10 Leveraging Performance Management Software
    Applications   245

11 Learning from Current Performance
    Management Practices   271

Further Reading   287

Index   289

Bernard Marr is a global authority and a best selling author on strategic performance management. In this capacity, he regularly advises leading companies, organizations and governments across the world, which makes him an acclaimed keynote speaker, researcher, consultant and teacher. Bernard Marr is acknowledged by the *CEO Journal* as one of today's leading business brains.

He has produced a number of seminal books, over 200 reports and articles on performance management and performance measurement, a number of Gartner Reports on Performance Management Software and the world's largest research studies on the topic. Organizations he has advised include Accenture, Astra Zeneca, the Bank of England, Barclays, BP, DHL, Fujitsu, Gartner, HSBC, Mars, the Ministry of Defence, the Home Office, Tetley, the Royal Air Force and Royal Dutch Shell.

Today, Bernard Marr heads up the Advanced Performance Institute (API) as its chief executive and director of research. He also holds a number of visiting professorships and editorial board memberships at many leading journals in the field. Prior to his role at the API, Bernard Marr held influential positions at the University of Cambridge and at Cranfield School of Management.

Bernard Marr's expert comments on performance management have been used in a range of high-profile publications, including in the *Financial Times*, *The Sunday Times*, *Financial Management*, the *CFO Magazine* and the *Wall Street Journal*. Bernard Marr can be contacted through e-mail at: bernard.marr@ ap-institute.com.

The **Advanced Performance Institute (API)** is the world's leading independent research and advisory organization specializing in organizational performance. The institute provides expert knowledge, research, consulting and training on concepts such as performance management, performance measurement and balanced scorecards.

The aim of the API is to provide today's performance-focused organizations with insights, advice and services that help them deliver superior performance. Customers of the API are wide ranging and include many of the world's leading blue chip companies as well as public sector organizations, governments and not-for-profit organizations across the globe.

Knowledge and Research

- The API conducts internationally recognized research with the aim of understanding and sharing the latest trends and best practices in the field of managing and measuring organizational performance.

- A wide selection of case studies, research reports, articles and management white papers are freely available to download from the API Web site.

Performance Management Audits

- Extensive research and implementation experience across the world puts the API into a perfect position to assess existing performance management approaches and compare them with current global best practices.
- Auditing and benchmarking solutions help organizations identify where they can improve their practices to get more value from their performance management and balanced scorecard initiatives.

Performance Management Consulting

- Based on the latest thinking, the institute can deliver a proven and tested process to designing performance management frameworks and initiatives.
- Perfected through real-life implementation experience across many industries, the institute is able to facilitate each step of the design process to ensure clearly articulated strategies, state-of-the-art strategic maps, meaningful performance indicators and aligned processes to ensure performance measures are communicated and used to inform day-to-day decision making and learning.

Performance Management Training

- The API provides training and coaching on any issues related to performance management. These are either offered as open enrolment training courses or delivered in customized workshops and training sessions within organizations.

For more information, please visit: www.ap-institute.com.

Preface

When I wrote my last book *Strategic Performance Management*, I was adamant that I could make it relevant to commercial organizations as well as government and not-for-profit organizations. However, over the past 3 years, I became increasingly convinced that there was a need for a more focused book dedicated specifically to the needs of government, public sector and not-for-profit organizations. My advisory work helping government, public sector and not-for-profit organizations across the globe with managing, measuring and delivering performance has taught me that there are many issues that are unique to those organizations. In addition, I have received many comments and e-mails from people in government and not-for-profit organizations who read *Strategic Performance Management*, asking me for case studies and examples of how the tools were applied outside the commercial business context. In hindsight, I could see that *Strategic Performance Management* is much more suitable for commercial organizations and that there was an increasing gap in the market for a practical and comprehensive book on managing and delivering performance in government, public sector and not-for-profit organizations.

Over a cup of coffee with my editor, I agreed to produce a government and not-for-profit edition of *Strategic Performance Management*. However, when I started writing the new edition, it became apparent that all chapters needed to be rewritten and new chapters to be added. As a result, I ended up with a completely new book, one that is entirely dedicated to the needs of government, public sector and not-for-profit organizations and that includes many new tools and techniques to cater for these needs.

All ideas, tools and templates provided in this book are both grounded in extensive research and fine-tuned through real-world applications. Without the extensive research, consulting and training work and the input and feedback from colleagues and clients, none of these ideas and tools would have materialized. I am grateful that I have had the opportunities to refine my knowledge during my time at Cambridge University, at Cranfield School of Management and now at the Advanced Performance Institute. I would like to thank all of my former colleagues who have contributed to my work and have provided knowledge and friendship.

In addition, I would like to thank the many federal and central government agencies; state and local government organizations; education organizations; police, fire or other emergency services; charities; courts; and national health care bodies that I have had the pleasure to work with. Without all these organizations, it would have been impossible for me to develop my insights and tools. I therefore would like to thank all of the many executives, managers and employees I have worked with over the years. By sharing their ideas and thoughts, all of them have helped me shape my thinking. Special thanks go to the governments of the United Kingdom, the United States and Australia and organizations such as the Ministry of Defence; the Bank of England; the European Patent Office; the Home Office; the Executive Office in Dubai; the New York City Police Department; the Department of Finance and Personnel; LPS; the Department of Culture, Arts and Leisure; the National Lottery Commission; Belfast City Council; Audit Scotland; the NHS, the Motor Neurone Disease Association; and the Royal Air Force.

Heading up the Advanced Performance Institute provides me with all the opportunities I need to develop and test new ideas, and I am indebted to my colleagues and the current fellows of the Advanced Performance Institute: David Teece, Rob Austin, Dean Spitzer, Péter Horváth, Klaus Moeller, Frank Buytendijk, Ian Shore, James Creelman, Leif Edvinsson, Marc André Marr, Mark Graham Brown and Paul Niven. Colleagues from other institutions and organizations who have been inspirational and significantly shaped my thinking for this book include Chris Argyris, Hans de Bruijn, Rob Grant, Douglas Hubbard, Chris Ittner and Bob Kaplan. I would also like to thank CIPFA (the Chartered Institute of Public Finance and Accountancy) for their support and collaboration, especially Brendan McCarron and the Performance Improvement Network (PIN) Team. Of course, there are so many other individuals who have also influenced my thinking, and I hope all of them know who they are and how much I have valued any input and dialogue over the years.

Finally, this book would never have been possible without the support from my wife, Claire, who gave up many of our precious evenings for me to write. My wonderful wife and our children, Sophia and James, are my clear priority in life and I am grateful that they support me. They provide so much inspiration to make the world a better place – and if this means helping organizations improve the way they manage and deliver performance, then this is a very small and humble, but hopefully useful, step into the right direction.

Managing and Delivering Performance

Managing and delivering performance is right at the center of any government, public sector or not-for-profit organization. Be it a federal or central government agency; a state or local government; an education organization; the police, fire or other emergency services; charities; courts; or national health care bodies, they all need to manage the effective and efficient delivery of their services.

In principle, performance management is very simple: First, you need to agree and clarify what matters in your organization; second, you need to collect the right management information to understand whether you are delivering performance in accordance with your plans; and third, you need to gain insights from the information, which in turn helps you deliver better performance going forward (see Fig. I.1). While in theory this is a simple and intuitive process, getting this right in our organizations seems not very simple at all.

In practice, I see that the execution of performance management is often very mechanistic and too number focused, preventing organizations from achieving the desired performance improvements. Instead of better performance, it often leads to frustration and decreasing performance, with perverse and dysfunctional behaviors such as target fixation, data manipulation, and cheating. Many government, public sector and not-for-profit organizations have created teams and departments who shed blood, sweat and tears to put performance management systems in place. Unfortunately, the result of these efforts is often just an increased administrative measurement burden and is very rarely producing new management insights, learning or performance improvements. I have written this book to change that situation and to provide people with a set of easy-to-follow tools, techniques and templates to create a truly performance driven organization in which managing and measuring what matters becomes everyone's everyday job and where it leads to real performance improvements.

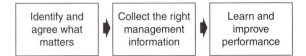

FIGURE I.1 Managing performance.

THE PERFORMANCE MANAGEMENT IMPERATIVE

Performance management has never been more critical for government, public sector and not-for-profit organizations than it is today. For many organizations across the world performance management is on the top of their management agenda. To help this process along, many governments have introduced legislations and frameworks to improve the management of performance in government and not-for-profit organizations. In the United States, for example, successive presidents have made strategic performance management part of their management agenda.[1] Back in 1993 the United States passed the Government Performance and Results Act that forces the head of each government agency to submit to the Office of Management and the Congress a strategic plan detailing the strategic aims and performance indicators. The key performance results are then aggregated into an executive branch management scorecard, which is published for everybody to see.[2] US President William J. Clinton, on signing the Government Performance and Results Act of 1993, said:

> … chart a course for every endeavor that we take the people's money for, see how well we are progressing, tell the public how we are doing, stop the things that don't work, and never stop improving the things that we think are worth investing in.

The UK Government is also taking performance management seriously and has designed a set of 90 so-called Best Value Performance Indicators to measure the performance of local authorities.[3] The UK Audit Commission assesses the performance of each local authority annually with the aim of helping them improve the services for their communities. This system is now being replaced by a Comprehensive Area Assessment with 198 national indicators at its core.[4] In Canada, the government has introduced a management framework for departments and agencies that includes a commitment to measurable improvements in client satisfaction.[5] In Australia, all government departments, agencies and business enterprises that deal with the public are required to develop customer service charters.[6]

In addition to these major initiatives in the United States, Canada, Australia and the United Kingdom, there are similar initiatives taking place in many other countries across the world including China, Sweden and the Netherlands. Other performance management and measurement initiatives focus more specifically on the police forces, health services, schools, universities and cities, among others. What most of these initiatives have in common is that they provide frameworks for managing and measuring performance, many prescribe predefined performance indicators with targets and several make the measurement data publicly available in, for example, league tables or performance scorecards.

The stated aims of these performance management initiatives tend to be improved performance with an emphasis on increased efficiency and effectiveness of service delivery and improved accountability to the public. While these aims make sense and the performance management approaches are generally well intended, many organizations in the public sector seem to approach performance management with an emphasis on collecting and reporting data that produces little insights, learning or improvement.

MANAGING ORGANIZATIONAL PERFORMANCE

I often compare organizational performance to a boat journey. At the beginning of any boat journey, you need to understand where the journey is heading. You wouldn't start a boat journey without a clearly defined destination, a map of how to get from where you are to where you want to be and the ability to keep track during your journey. On a boat, everybody understands the destination and the route and they are clear about their role in making the journey a success. Once the boat has left the harbor and you are sailing in the middle of the ocean, you need to assess whether the boat is still on course or whether corrective actions have to be taken. Without a clear route map and appropriate instruments that help you understand where you are, you are unable to establish whether you are still on route, and you are unable to understand whether any actions you have taken have brought you closer to your destination or not. Without a clear plan and the necessary information, it is impossible to navigate the boat to success, and the same is true for an organization.

When I talk about performance management in this book, I talk about creating an environment in which organizational performance becomes everyone's everyday job. In such an environment, everybody in an organization clearly understands the strategic priorities and accepts responsibility for the delivery and continuous improvement of performance. Employees intuitively use performance information to inform decision making at all organizational levels, and not merely to put them into reports that no one really cares about.

By strategic performance management I mean the organizational approach to clarify, assess, implement and continuously improve the organizational strategy and its execution. It encompasses strategic frameworks, performance indicators, methodologies and processes that help organizations with the formulation of their strategy and enable employees to gain relevant insights, which allows them to make better-informed decisions and learn.

It therefore goes far beyond the narrow definitions of performance management as just collecting and reporting data and it is more than just people management. Strategic performance management is about identifying, measuring and then managing what matters in order to improve the effectiveness, efficiency and overall performance of an organization.

If I compare this definition of performance management with the practice I am seeing in most government, public sector and not-for-profit organizations today then the picture I am seeing is quite far removed from what I have described as good performance management. The picture I am seeing day in and day out is that:

- The majority of organizations spend too little time clarifying and agreeing strategy. In many organizations, the annual planning process is a big waste

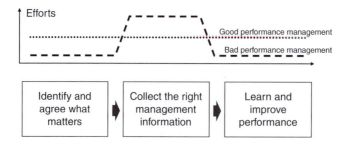

FIGURE I.2   Right and wrong emphasis.

of everyone's time as they end up creating a vague and seemingly incoherent business plan that is either a rewrite of last year's plan or a bottom-up aggregation of everything anyone is already doing.

- Most organizations seem to spend too much time measuring everything that is easy to measure. A huge chunk of management time is taken up with collecting meaningless and irrelevant performance indicators, which means organizations end up drowning in data while thirsting for information.
- Organizations don't spend enough effort ensuring the performance data is turned into meaningful insights and learning. The majority of organizations I see just stick the performance data into reports, which are e-mailed to everybody hoping that by magic this is somehow helping to improve performance.

In Fig. I.2 the dashed line in the graph indicates this scenario, where too little time is spent clarifying and agreeing strategy, too much time is spend collecting performance data and again too little time is spend doing anything meaningful with the data. Instead, what we need to do is balance this picture out a little (see dotted line). Organizations need to spend more time and effort on creating and agreeing a clearly articulated and coherent strategy, then they need to focus only on collecting relevant and meaningful performance information based on their strategic priorities, which then allows them to spend more time and efforts on using the performance information to extract management insights and learning in order to improve performance.

IDENTIFY AND AGREE WHAT MATTERS

If any organization wants to reach its goals, it must first know what they are, so everyone can pull in the same direction. Unfortunately, government, public sector and not-for-profit organizations are notoriously bad at clarifying their strategy. One of the essential premises of this book is that strategy formulation (or, more commonly, reformulation) is an essential prerequisite for successful performance management. Strategy development is a layered process that should start with an analysis of the environment in which the organization operates

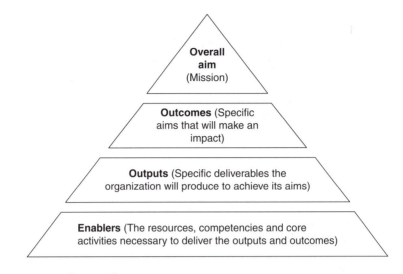

FIGURE I.3   Elements of strategy.

including an assessment of the key stakeholders and their requirements. This allows organizations to clarify their outcome objectives and output deliverables in the form of a clear value proposition. However, that is not enough.

Organizations have realized that they need to ensure they have the appropriate internal competencies and resources to deliver the outcomes and outputs. Having ambitious objectives without the underlying ability to deliver them will not lead to success. One strategy thinker brought this to a point when he said that 'opportunism without competence is a path to fairyland'.[7] In order to create a more complete picture of the strategy, an organization needs to clarify its overall aims, outcomes, outputs as well as the enablers of performance (see Fig. I.3).

It is important to understand that these various components of strategy are interdependent and sit in a cause-and-effect relationship with each other. An analogy might help to illustrate how these fit together. Think of the organization as a tree (see Fig. I.4).[8] Its foliage is how it presents itself to the external world and its fruits (say, apples) are the products or services it offers to its customers and stakeholders. The major branches of the tree represent the set of departments and service units in the portfolio of an organization. The tree's trunk represents the core activities that give it its strength and hold the tree up; they ensure the tree can deliver its apples (products and services). The tree's hidden roots, on the other hand, represent the enablers of performance; that is the tangible and intangible resources it needs to have in place in order to provide the sustenance it requires to grow the apples that customers and stakeholders require. The trunk therefore provides the channel leveraging the resources to create value. Similar to organizations, all trees are made up of the same elements and

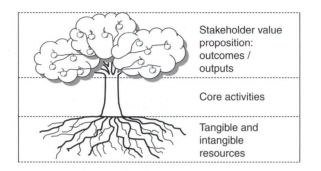

FIGURE I.4   Tree analogy.

share the same biological processes of photosynthesis and nutrient extraction, but the shape of the trees and their fruits differ widely.[9]

What apple trees cannot do, of course, and what organizations usually must do is to grow a blend of red and green apples at the same time in different quantities according to the demand for each type. Nevertheless, defending this slight snag with the usefulness of the analogy, the owner of an orchard can plant a mixture of trees that provide a supply of both green and red apples. The organization might then perhaps be better considered as an orchard rather than a single tree.

The purpose of this analogy is to highlight the point that organizations create value not only by understanding their outcome objectives, but also by having a deep understanding of the nature of their core activities and resources required to deliver them. Organizations embarking on any strategic performance management initiative, therefore, need to analyze information not only about their external markets but also about their internal resources. The danger here is that organizations develop a one-sided view of strategy that does not connect their stakeholder value proposition with their internal activities and resource infrastructure.

Once organizations understand the different elements of their strategy, they can map them into a strategic map that illustrates how the different elements work together to create value. Such a visual representation of the organization strategy is one of the most important components of successful performance management as it allows communicating the strategic plan on a single piece of paper. This integrated and coherent strategy is then the starting point for organizational alignment and for any performance indicators.

In Part I of this book, I outline how you clarify and agree a strategy by discussing each layer of the strategy development process in detail (see Fig. I.5). For each step I provide a number of tools, templates and practical examples to ensure you create a well-defined strategy as a solid foundation for good performance management. Throughout Part I many real-life case studies help to illustrate how these tools can be applied in practice.

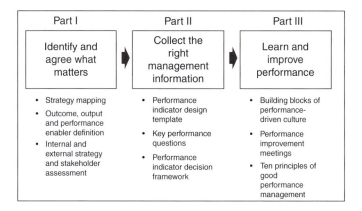

FIGURE I.5   Structure of the book.

COLLECT THE RIGHT MANAGEMENT INFORMATION

Performance indicators are vital tools for our organizations. Going back to the boat analogy, without the right management information we can understand nothing and know nothing. At the same time, in order to be of value, performance indicators must help us assess the things that matter the most, and not merely those that are easy to measure. We need to measure those things that are directly linked to the strategic objectives of the organization and we need to understand that it is not just about numbers. We have become so obsessed with quantifying that we sometimes forget that performance information is only any good if it helps us to gain new insights. And in order to do this we have to supplement numbers with words and commentary.

Performance indicators only help us understand whether we are on the right track or not if they have real information value. Indicators have to help us answer our burning and unanswered questions and therefore help us to make better decisions. It is therefore critical that we identify and articulate what these questions are before we start collecting any management information. The danger here is that organizations don't link their indicators to the strategy of the organization and that they attempt to quantify the unquantifiable or measure everything that is easy to measure without focusing on the relevant and meaningful indicators in order to use them for strategic decision making and learning. This quickly leads to institutionalized and bureaucratized systems of measurement. Instead, organizations need to assess what they value rather than value what is assessed.

In Part II of this book, I take a detailed look at measuring performance and provide practical guidance on some of the key pitfalls that we need to avoid when it comes to measurement (see Fig. I.5). I introduce the new concept of key performance questions and outline a number of templates and frameworks to develop relevant and meaningful performance indicators. These will help you to collect management information for anything in your organization – even the most intangible and seemingly immeasurable aspects of performance. I also provide a range of innovative performance indicators developed by government, public sector and not-for-profit organization.

LEARN AND IMPROVE PERFORMANCE

Once we have defined the strategy and derived relevant performance information, we need to use the information for evidence in our decision making. Too many organizations believe that once they have collected performance indicators and put them into spreadsheets or reports, the work is done. Measurement is often seen as an end in itself. Instead, it is a means to an end and only the beginning of any improvement journey.

For any learning to take place, we need to develop the right culture and attitude towards performance improvement. Many organizations still apply a machine-like command-and-control model in which they use measures to control people's behavior. This is not only inhumane but also goes against the increasing need for organizations to be more adaptive to ever changing stakeholder needs. If we therefore use measures to treat people like machines and try to control them like robots, we will just get what we measure, which is not what we want! Dumbing people down with metrics is the wrong form of accountability.

If we fail to acknowledge the important role of organizational culture in making performance management a success, then we are likely to get dysfunctional behaviors and gaming of measures. People might be hitting the target but missing the point. Instead, we need to create a performance-driven culture with an enabled learning environment. Here, learning and improvement take center stage. In such an environment, performance measurement empowers people to make better-informed decisions and to become accountable for performance delivery. In this culture, performance measurement and management is not done to people, instead everyone is truly engaged in and in charge of performance management.

The danger here is that organizations either collect too many irrelevant measures only for reporting purposes or they use measures to gain additional top-down control; both situations mean that measures are unlikely to be used for any strategic decision making or learning.

In Part III of this book, I look at how we create the right organizational culture and the right processes to turn our data into information and that information into learning and performance improvement (see Fig. I.5). I discuss the key components of a performance-driven culture and how we leverage performance management software applications to engage everyone in managing performance, and I outline 10 principles of good performance management.

REFERENCES AND ENDNOTES

1. See the President's Management Agenda: www.whitehouse.gov/results/agenda/faq.html
2. See the President's Management Agenda Scorecard: www.whitehouse.gov/results/agenda/scorecard.html
3. For more information on Best Value Indicators please see www.bvpi.gov.uk
4. See Comprehensive Performance Assessment: www.audit-commission.gov.uk/cpa/
5. See Service Improvement Initiative as part of the Results for Canada management framework: www.tbs-sct.gc.ca
6. See Putting Service First – Principles for Developing a Service Charter: http://parlinfoweb.aph.gov.au/piweb/
7. Andrews, K. R. (1971). *The Concept of Corporate Strategy*. Dow Jones-Irwin, Homewood, IL.
8. This analogy has been used on various occasions, one of the most convincing was by Prahalad, C. K. and Hamel, G. (1990). The Core Competence of the Corporation. *Harvard Business Review*, 68(3), 79. However, tree diagrams can be traced back to the third-century Syrian philosopher's diagram, named after its developer 'Tree of Porphyry', based upon the work of Aristotle
9. Collis, D. J. and Montgomery, C. A. (1997). *Corporate Strategy – Resources and the Scope of the Firm,* McGraw-Hill, Boston, MA. p. 130.

Part I

Identifying and Agreeing What Matters

Identifying and agreeing on strategic objectives and priorities is essential in any government and not-for-profit organization. However, I see many of these organizations struggling to agree on strategic priorities. One reason for this is the fact that they often have different external stakeholders with differing, and sometimes opposing, external demands and requirements. As a consequence, organizations have difficulties answering the question of why they exist and struggle to agree on their key output and outcome deliverables. Is your organization here to serve the government, the citizens or a specific community? If you or anybody in your organization finds it difficult to answer this or if they come up with different answers, then there is a problem. To use a nautical analogy – how can you expect everyone in the boat to help row the boat forward if no one is clear about which direction the boat should be heading? It is obvious that in order to provide value to customers and deliver best performance, government sector and not-for-profit organizations need to agree on strategic deliverables and then create plans that will enable them to deliver these deliverables in the most effective and efficient manner.

This brings me to the next problem. What I see quite often is that even if government and not-for-profit organizations have agreed on a list of output and outcome deliverables, they often fail to link them to internal competencies and resources required to deliver these outputs and outcomes. This means that lack of competence or resource limitations are often ignored when designing public service and not-for-profit strategies. Strategic objectives without the capacity, resources and competencies to deliver are no more than wishful thinking. Any good strategy therefore brings together the outcomes and external value proposition with the internal competencies and resources requited to deliver them.

I have seen so many business plans and strategies, especially in public service and not-for-profit organizations, that are not worth the paper they are written on. The exercise of creating a strategic plan is far too often just an administrative burden – something you have to do but don't want to do. The result is that instead of being powerful documents that clearly lay out priorities and objectives, they often are a wish list of strategic objectives, regularly based on the objectives of the previous years, written up in a 35-page document that really no one ever reads or understands.

Managing and delivering performance in the government and not-for-profit sector is about engaging everyone in the strategy and its execution so that organizational performance becomes everyone's everyday job. The starting point for good performance management is therefore a shared understanding and clarification of the strategic context of the organization. We cannot expect people to understand and implement our strategy if they don't know what the strategy is.

Ensuring that everybody knows where the journey is heading before setting off might seem intuitive to the managers and executives involved. However, my experience has taught me differently. Far too often, organizations embark on their journey to manage performance without clarifying their strategy. Ignoring the thorough examination of the external and internal contexts of an organization's strategy is a mistake that we tend to make time and again. And even if organizations do understand the strategic context, it is often a one-sided view where they look at either external opportunities or internal competencies. The main reason for this is that we are too often deeply submerged in the everyday microdetail of the organization's workings. However, if we want to make strategic performance management a success then we need to come up to the surface, take a deep breath and have a realistic look at where we are.

While some managers and executives may feel that they clearly understand their organization's strategy, my experience is that this understanding is often their interpretation of the strategy and that others have a significantly different interpretation of what the strategy is. Developing a common and shared understanding of the organization's direction is one of the most valuable and rewarding exercises. This shared understanding can then be translated into a visual and narrative summary of the organizational business model. A so-called 'value creation map' can be created to bring together on one piece of paper the key components of the organization's strategy, namely the stakeholder value proposition and the core activities required to deliver the value proposition, as well as the key resources (tangible and intangible) that underlie the core activities. This is then accompanied by a brief one-page narrative summary of this business model called the 'value creation narrative'.

The value creation map and value creation narrative describe the business model and therefore create a shared understanding of the strategy. This in turn helps to create a common purpose, a shared identity and a sense of community. This understanding of strategy can then be used to align the organizational activities, allow organizations to manage their relevant risks and create the right organizational structure and governance, before it can guide the development of meaningful and relevant performance indicators, which can then be used to challenge and refine the business model and its assumptions.

Too many performance management approaches assume that the strategy and business models are well understood by everyone in the organization. From my experience, this is not always the case and this is often a key contributing factor to the failing of performance management initiatives. The following chapters bring different components of strategic management together to form a template for what needs to be addressed in order to define the strategy and the business plan. Depending on how well your organization's strategy is defined and understood, you can select the appropriate starting point. For many government and not-for-profit organizations or many business units within them, the external context is somewhat dictated by the overall organizational purpose.

For example, the police forces have to continue to play their role in crime prevention as outlined by their governments. In the same way, central government departments have to perform their assigned roles, and if you are working for a cancer charity then your main objective will always be to provide funding and care for people affected by cancer. This means the overall direction is set by the purpose and requirements of external stakeholders. The government and not-for-profit organizations have not got the same freedom that many commercial organizations enjoy, and therefore, they can rarely decide to move into a completely different overall purpose.

Even though there are some clear boundaries set in terms of overall purpose, the finer details still need to be clarified. For example, one of the overall purposes of the police force might be crime prevention; however, this does not clarify current priorities. It is still important to identify and agree on what to focus on, especially in a world with resource limitations. For example, the police might decide to focus on reducing more serious violence and increasing community confidence, or they might focus on the increased threat posed by violent extremists to communities. The point I am trying to make here is that even though there might be a predefined overall purpose, the strategic options are still endless and agreement and clarity need to be achieved in order to make any strategy actionable.

Once the strategy has been agreed and mapped, it is important to align organizational activities and projects with this strategy. This is a crucial link that is often somewhat ignored. The organizations need to ensure that they have the right projects and initiatives in place to deliver their different strategic objectives. This might seem trivial but again my experience has taught me differently. The organizations are full of projects and initiatives that are not necessarily linked to the delivery of the newly defined strategy. It is therefore important to map the existing activities onto the strategy to ensure it can and will be delivered. Initiatives and projects make the strategy real; without closely aligned initiatives and projects the strategy will never be delivered. This brings me to a related problem: the budgeting processes. Initiatives and projects require resources and funding, which means that the budgeting process should be driven by your strategic objectives and the necessary activities required to deliver them. Unfortunately, the budgeting processes I have observed in many public service and not-for-profit organizations are over and over again completely detached from the strategy process and often highly dysfunctional and arbitrary. Again, how can anybody expect the delivery of the strategy if the funding decisions are made in ways where budgets for business units are either raised or cut by a certain percentage figure without any considerations of the strategic needs?

The other element of good strategic performance management is the mitigation of risks. Risk management is high up on the agenda of most government and not-for-profit organizations. The problem is that risk management is not often aligned with the strategy. If our strategy identifies the crucial deliverables and enablers of performance, then we have to consider these elements in our risk management activities.

In this part of the book, I outline the latest tools that have been designed to enable organizations to identify

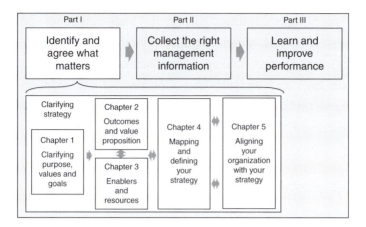

FIGURE P1.1   Overview.

and agree on what matters. Figure P1.1 outlines the structure and chapters in this part of the book. I will start by looking at the overall purpose, values and goals which define the boundary conditions that delimit the confines in which an organization operates. In most cases, these are already defined. Chapter 1 describes what they are and their role in the organization. Chapter 2 looks at how to define the outcomes and the overall value proposition of the organization before I discuss in Chapter 3 the enablers and resources required to deliver the strategy. Chapter 4 then describes how these insights can be translated into a business model, visually represented in a value creation map and described in a value creation narrative. If you believe your organization already has clearly articulated outcome objectives and deliverables, then you might go straight to the internal analysis (Chapter 3). If you believe both the external and internal contexts are understood and objectives have been clearly defined (likely to be quite rare), you can go straight to the mapping and narrating your value creation (Chapter 4).

Once I have provided the tools and techniques for creating a truly integrated and cohesive strategy, I will move to Chapter 5 where I will discuss the alignment of activities and budgets, the alignment of risk management, and the alignment of organizational structure and governance.

Chapter 1

Clarifying Purpose, Goals and Values

All organizations need to adapt over time – to changes in their overall direction, to regulatory demands, to changing stakeholder wants and needs, or to evolving and changing internal competencies. Nevertheless, some aspects of 'what the enterprise is there to do and how it will go about doing it' remain relatively constant through time. Statements pertaining to overall purpose, visionary goals and core values are created by organizations in order to provide the overall guiding principles for their strategic thinking and their employees' behavior. These are usually established when an organization is founded; however, they sometimes, but infrequently, need to be changed when the organization takes a fundamental change of direction, such as in a situation in which government departments are merged and where perhaps conflicting visions need to be harmonized. Essentially, what we are seeking here is the 'glue' that holds the whole organization together over a fairly long period of time and sets the general boundaries within which an organization operates.

However, this should not be confused with the frequently changing platitudes iterated by successive chief executives, which purport to be visionary mission statements but say much the same things that other organizations in the same field expound because it is almost customary to emphasize certain characteristics. Usually, such statements are, in fact, more like current strategic ambitions. Here, we are looking for sound, long-lasting and differentiated definitions of the very raison d'être of the enterprise. The questions I address in this chapter include the following:

- What are strategic boundary conditions?
- What is the core purpose?
- What are the visionary goals?
- What are core values?
- How do boundary conditions set limits on the forward strategy?

In order to assess the strategic activities and the existing boundaries for your organization, it is important to assess the following three essential components:

- The fundamental *purpose* of the enterprise
- The long-term visionary *goals* that the enterprise will pursue
- The core *values* that the enterprise commits to on the way to delivering its purpose and achieving its goals

CORE PURPOSE AND MISSION STATEMENTS

The core purpose of an organization is the overall reason for which it exists. This is often described in so-called purpose or mission statements, which are precise descriptions of why an organization exists and what it does. It is normally expressed in brief, enduring and often loosely idealistic terms that nevertheless provide an overriding direction and clarification of its ambitions. Mission statements serve as ongoing guides without a time frame, which can remain the same for decades if crafted well.

In case of commercial enterprises, the primary purpose is to generate profits and uphold the interests of shareholders. However, many commercial companies take it a lot further as they realize that generating profits is not what makes their employees get out of bed in the morning. We all want to buy into a mission to deliver some greater good or greater purpose. Below I have listed three examples from Merck, Microsoft and Google that illustrate such greater ambitions.

Merck:

Provide society with superior products and services by developing innovations and solutions that improve the quality of life and satisfy customer needs, and to provide employees with meaningful work and advancement opportunities, and investors with a superior rate of return.

Microsoft:

To enable people and businesses throughout the world to realize their full potential.

Google:

Organize the world's information and make it universally accessible and useful.

I feel that this need for a 'worthwhile' organizational purpose is even greater in the government and not-for-profit sector. My experience is that people generally join charitable organizations and government departments because they want to contribute to a good cause and provide a meaningful public service. While public service and not-for-profit organizations do not share the unifying purpose of generating profits and shareholder value, they do have a strong public service imperative. Clarifying the fundamental purpose of an organization can be a powerful means, if done well, to drive an entire organization from top to bottom.[1]

Below I have listed three examples of short and snappy mission statements from the UK Ministry of Defence, the US Department of State and Harvard University's Business School, along with three more elaborative mission or purpose statements from the City of Coronado Fire Department in San Diego Bay in California, The Canadian Banting Research Foundation and the Australian Government's Treasury.

UK Ministry of Defence:

A force for good in the world.

US Department of State:

Create a more secure, democratic, and prosperous world for the benefit of the American people and the international community.

Harvard Business School:

We educate leaders to make a difference in the world.

City of Coronado Fire Department:

The mission of the Coronado Fire Services Department is to prevent fires from occurring and, when they do occur, to utilize resources efficiently and effectively to protect life and minimize property damage; to provide emergency medical services including patient transportation; and to prepare the community, through plans and education, for natural disasters.

The Banting Research Foundation:

The Banting Research Foundation supports promising young medical researchers across Canada, doing a broad range of medical research and who are setting up their first independent laboratory. Unlike many other foundations, it does not focus on a specific disease.

The Australian Government's Treasury:

Treasury's mission is to improve the wellbeing of the Australian people by providing sound and timely advice to the Government, based on objective and thorough analysis of options, and by assisting Treasury Ministers in the administration of their responsibilities and the implementation of Government decisions.

VISIONARY GOALS AND VISION STATEMENTS

The visionary goals outline what an organization aims to achieve in the long run. They describe desired outcomes that inspire and energize and help people to create a mental picture of the target. These are usually expressed in so-called vision statements that outline what an organization wants to be or wants to achieve. The visionary longer-term goals of the organization are clearly not only for the next year or two, but are milestones that the enterprise will endeavor to achieve within perhaps 10 years' time.

A vision statement is essentially a declaration of the organization's medium-term ambitions, either quantitative or qualitative goals, which are far beyond current performance levels.[2] For example, these might include 'stretch goals' such as becoming the dominant player in a particular field, reaching a certain size, becoming the best at something, beating a particular competitor, or becoming a role model in a specific sector and so on. In the early 1900s, Ford Motor Company set out to 'democratize the automobile' and, in the early 1950s, Sony's visionary goal was to 'become the company most known for changing the worldwide poor-quality image of Japanese products'. Boeing and Wal-Mart set more measurable visions. In 1950, Boeing's vision was to 'become the dominant player in commercial aircraft and bring the world into the jet age' and, in 1990, Wal-Mart's vision was to 'become a $125 billion company by the year 2000'.

The common theme is that they have a challenge that is not easily achievable. When, eventually, these challenges are successfully met, they have to be replaced with a new challenge in order for the organization to rejuvenate itself. It is unfortunate that many public services and not-for profit organizations omit doing this and hence become vulnerable to complacency. Below are some illustrative examples of public service and not-for-profit vision statements from the John F. Kennedy Memorial Hospital in Indio, CA, USA; the Cleveland Police, UK; the Eden Baptist Church in Cambridge, UK; and the Indiana University Kokomo, IN, USA.

John F. Kennedy Memorial Hospital:

> JFK Memorial Hospital's vision is to become the first choice for health and medical services for residents of the Eastern Coachella Valley.

Cleveland Police:

> By 2010 Cleveland Police will be at the forefront of modern policing, driving forward problem solving police work, through close co-operation with our partners, in response to the real needs of the communities we serve.

Eden Baptist Church:

> We desire to be a family, rooted in the Word of God and prayer, whose love for Christ and for one another overflows in joyful sharing of the gospel with all, especially with those friends, neighbors, students and internationals whom God has given us a unique opportunity to reach.

Indiana University Kokomo:

> Indiana University Kokomo aspires to become a regional institution of first choice recognized for providing critical opportunities for student success; acknowledged as a primary and engaged community resource; and valued as a campus where there are faculty, students, and professional staff active in research, creative work, and other scholarly activity.

CORE VALUES AND VALUE STATEMENTS

Core values should reflect the deeply held values that the organization espouses and should be totally independent of strategic priorities or topical management fads. Values can set an enterprise apart from the competition by clarifying its identity, limiting its strategic and operational freedom and constraining the behavior of its people.[3] Core values can be articulated in so-called value statements that set certain boundaries on the behavior of people in the organization. Core values outline the principles with which people in an organization interact with the world within and around them. The core values contained in a value statement should be few in number (typically not more than five to seven, so that they are memorable), but they can be expressed in lengthier prose than the mission and vision statements.

Some organizations take the prescription of values and behaviors a little far when they create those corporate 'good behavior' booklets that contain mounds of rules and behaviors. I have, in fact, seen that one organization had 123 rules for its people to follow every day and another had 144 rules of 'leadership impera-tives' that its managers should 'live and breathe' (not to mention the further 82 principles that managers must *not* do). The problem is that a long list of values and behaviors becomes meaningless as no one can remember them.

Let's look at some illustrative examples. The UK Home Secretary has outlined a number of shared core values that policing must always be rooted in. These are as follows:

> Fairness and impartiality, integrity, freedom from corruption, respect for liberty and com-passion. It must be free from racism, serve all communities equally, and be committed to our individual protection and our common well-being.[4]

A well-known example of value statements comes from the commercial firm Johnson&Johnson. The giant US pharmaceuticals and healthcare firm has, and continues to maintain, its core values in what it calls its 'credo'. This has been in place for over 60 years, with only very minor clarifica-tions introduced over time. First created in 1943 by General Robert Wood Johnson, it is a one-page document that sets out the firm's 'industrial phi-losophy' as to the corporation's responsibility to its various stakeholders. Sometimes seen as controversial, it puts customers first and shareholders last in its list of priorities. The company legitimately claims that its employ-ees have made countless decisions that were inspired by the philosophy embodied in the credo, and that these have succeeded in enhancing the company's reputation (not least during the company's well-known Tylenol product recalls in the 1980s). The full text of this philosophy is reproduced below:

Johnson&Johnson's credo:

> We believe our first responsibility is to the doctors, nurses and patients,
> to mothers and fathers and all others who use our products and services.
> In meeting their needs everything we do must be of high quality.
> We must constantly strive to reduce our costs
> in order to maintain reasonable prices.
> Customers' orders must be serviced promptly and accurately.
> Our suppliers and distributors must have an opportunity
> to make a fair profit.
>
> We are responsible to our employees,
> the men and women who work with us throughout the world.
> Everyone must be considered as an individual.
> We must respect their dignity and recognize their merit.
> They must have a sense of security in their jobs.
> Compensation must be fair and adequate,
> and working conditions clean, orderly and safe.
> We must be mindful of ways to help our employees fulfill
> their family responsibilities.

Employees must feel free to make suggestions and complaints.
There must be equal opportunity for employment, development
and advancement for those qualified.
We must provide competent management,
and their actions must be just and ethical.

We are responsible to the communities in which we live and work
and to the world community as well.
We must be good citizens – support good works and charities
and bear our fair share of taxes.
We must encourage civic improvements and better health and education.
We must maintain in good order
the property we are privileged to use,
protecting the environment and natural resources.

Our final responsibility is to our stockholders.
Business must make a sound profit.
We must experiment with new ideas.
Research must be carried on, innovative programs developed
and mistakes paid for.
New equipment must be purchased, new facilities provided
and new products launched.
Reserves must be created to provide for adverse times.
When we operate according to these principles,
the stockholders should realize a fair return.

The Canadian Cancer Society provides another illustrative example of values that serve as guidelines for the conduct and behavior as the society works toward their vision of creating a world where no Canadian fears cancer:

- Quality: Our focus is on the people we serve (cancer patients, their families, donors and the public) and we will strive for excellence through evaluation and continuous improvement.
- Caring: We are committed to serving with empathy and compassion.
- Integrity: We are committed to act in an ethical, honest manner.
- Respect: We believe that all people should be treated with consideration and dignity. We cherish diversity.
- Responsiveness: We strive to be accessible, flexible and transparent, and to demonstrate a sense of urgency in our resolve and decision-making.
- Accountability: We are committed to measuring, achieving and reporting results, and to using donor dollars wisely.
- Teamwork: We are committed to effective partnerships between volunteers and staff, and we seek opportunities to form alliances with others.

However, there are dangers lurking here too, the most pressing one being adopting blandly nice ideals that fail to differentiate an organization from its competitors. Patrick Lencioni, a leading thinker in this field, says:

Consider the motherhood-and-apple-pie values that appear in so many companies' values statements – integrity, teamwork, ethics, quality, customer satisfaction, and innovation. In fact, 55% of all Fortune 100 companies claim integrity is a core value, 49% espouse customer satisfaction, and 40% tout teamwork. While these are inarguably good qualities, such terms hardly provide a distinct blueprint for employee behavior. Cookie-cutter values don't set a company apart from competitors; they make it fade into the crowd.[2]

The art of designing good value statements is to make them real and relevant. If they do not reflect your organization and are just a generic subset of values that every other organization uses, then they become meaningless. However, if you are able to create a list of values that reflect your organization and its people and if you are able to distill them down to less than seven meaningful statements, then they can become powerful guiding principles for setting behavioral standards in your organization. I feel that the entertainment company Walt Disney is a great example of how an organization can create a concise set of unique, meaningful and relevant values. Their values are as follows:

- No cynicism
- Nurturing and promulgation of 'wholesome American values'
- Creativity, dreams and imagination
- Fanatical attention to consistency and detail
- Preservation and control of the Disney 'magic'

WHY DO PURPOSE, GOALS AND VALUES MATTER?

Purpose, goals and values are important factors because they set the *boundary conditions* for an organization's forward strategy (see Fig. 1.1). They need to be taken into account when defining any strategy. But as Jim Collins and Jerry Porras, authors of *Built to Last: Successful Habits of Visionary Companies*, point out:

> Many executives thrash about with mission statements and vision statements. Unfortunately, most of those statements turn out to be a muddled stew of values, goals, purposes, philosophies, beliefs, aspirations, norms, strategies, practices, and descriptions. They are usually a boring, confusing, structurally unsound stream of words that evoke the response 'True, but who cares?'[5]

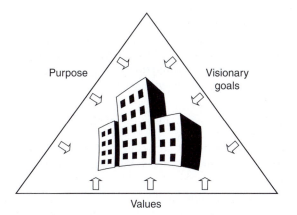

FIGURE I.I   Purpose, goals and values as boundary conditions.

In fact, many of the mission statements of government and not-for-profit organizations I have visited seem to come straight out of the games section of the dilbert.com Web site, which provides the wonderful humorous facility of a random 'mission statement generator'. How about these: 'The customer can count on us to collaboratively utilize world-class intellectual capital and professionally administrate performance-based resources to exceed customer expectations' or 'We envision to professionally customize emerging data so that we may endeavor to interactively build economically sound resources to stay competitive in tomorrow's world'. These admirably satirize the sheer vacuousness of the majority of the mission statements.

What to do if your organization is one of those? If these components are not properly in place already, then the organization in question has no guiding beacon or principles within which to operate. This means that it has no cohesive view of its overriding objectives, no long-term policies, no view of how it should interact with its various stakeholders and no guidance as to how its employees should behave. Where that is the case, moving forward with strategic performance management decisions is by all means possible but should be approached with great care. I have worked with various organizations where these guiding principles were in place but not made explicit, or others where these principles were developed and then forgotten or buried somewhere in the organizational databases. It usually makes sense to dig them up and run a few simple tests:

- Do they make sense? (It is amazing how many do not!)
- Are they up to date?
- Are they relevant?
- Are they easy to understand?

If the answers to these questions are yes, then it might be time to revitalize these and take them into consideration for the forward strategy. If the answers to these questions are no or if your organization does not have any explicit purpose, goals and values, then it might be time to create or revise them.

The output of this initial stage should be a clarification of the boundary conditions within which any new strategy will be set. It simply sets out the essential basis for the enterprise to move forward without harming the central premises to which it aspires. It is important to be clear about these fundamental basic building blocks before moving on to design a strategy.

SUMMARY

- The purpose, visionary goals and values for government and not-for-profit organizations have been discussed as they are important in defining the forward strategy.
- If crafted well, they set the *boundary conditions* for an organization's forward strategy. However, they are rarely designed well and therefore are not as guiding and powerful as they could be.

- Together, they provide a cohesive view of an organization's overriding objectives, long-term goals and a view of how it should interact with its various stakeholders and how its employees should behave.
- The *core purpose* is expressed in a *mission statement* that articulates what the organization is at present and describes why an organization exists and what it does. Mission statements serve as ongoing guides without a time frame and can therefore remain the same for a long time.
- The *visionary goal(s)* are expressed in a *vision statement* that outlines what an organization wants to be. It concentrates on the future, creates a mental picture of *a specific medium-term* target and is a source of inspiration.
- The *core values* are expressed in a *value statement* that articulates the desired behavior of people in the organization. It outlines the principles with which people in an organization interact with the world within and around them.
- These boundary conditions form the first part of strategy definition but need to be integrated with the external and internal analysis discussed in the following two chapters.

REFERENCES AND ENDNOTES

1. See, for example, Osborne, D. and Gaebler, T. (1992). *Reinventing Government*. Jossey-Bass, San Francisco
2. Referred to, in late 1990s-speak, as 'Big, Hairy, Audacious Goals' in Collins, J. C. and Porras, J. I. (1996). Building Your Company's Vision. *Harvard Business Review*, Sept.–Oct., 126–148
3. Lencioni, P. M. (2002). Making Your Values Mean Something. *Harvard Business Review*, July, 113–117.
4. Common values for the Police Service of England and Wales, Home Office, http://police.homeoffice.gov.uk
5. Collins, J. C. and Porras, J. I. (1996). Building Your Company's Vision. *Harvard Business Review*, Sept.–Oct., 126–148.

Chapter 2

Understanding Your Outcomes, Outputs and Value Proposition

Outcomes and value propositions define what the organization is planning to deliver and whom it is planning to deliver it to. This means organizations have to take a hard look beyond the boundaries of the organization and make informed choices about the value and outcome it intents to deliver. This involves an external analysis of both the macroenvironment in which the organization operates and the microenvironment in which it competes with other providers of similar services or products. The key outcome from this external analysis is the stakeholder value proposition – basically an answer to the questions who your key stakeholders are and what you are planning to deliver to them.

To follow my tree (or orchard) analogy and compare your organization with this metaphoric apple tree, the questions I believe you should try to answer include (see Fig. 2.1):

- Who are our key stakeholders and what do they expect from our organization?
- What kind of service or product attributes do our stakeholders require?
- What changes, challenges and discontinuities are there in the external environment?
- What are the different competitive forces in our sector or market?
- How would changes in the overall macroenvironment (economic, social, political, etc.) impact our value proposition?

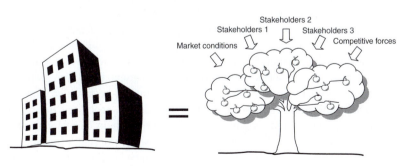

FIGURE 2.1   External environment.

One would expect most organizations to understand this part of their strategic context relatively well. However, I find many government and not-for-profit organizations struggling with it. Part of the added difficulty for government and not-for-profit organizations is that they operate in a multiple values environment where they have different external stakeholders who might have different or even opposing demands. In addition, the outcomes often depend on efforts of third parties.[1] Professor de Bruin has looked into this aspect in more detail and provides the following examples. The objectives of a school, for instance, are to make its pupils perform well in exams as well as to provide a good educational climate (multiple values), and the performance and learning of pupils depend somewhat on the level of stimulation they receive at home (coproduction). I can think of many public service and not-for-profit organizations where this is also the case. For example, the police forces face a multiple stakeholder environment: some neighborhoods might demand good levels of police presence on the street and in the area they live in; at the same time, key pressure groups might demand that the police deal with specific organized crime and the government might impose a set of different crime reduction targets despite resource constraints. The performance, on the other hand, as defined by the outcomes (e.g. crime reduction) depends on other external factors such as demographics, general crime levels, policies, culture, etc., which makes it sometimes hard to reach an agreement on the stakeholder value proposition.

Government and not-for-profit organizations also find the process of defining strategic outcomes and value propositions somewhat foreign. They sometimes feel that by clearly articulating their intended outcomes they somehow limit their freedom. Instead many seem to feel much more comfortable leaving their objectives vague and open so that they have the freedom to deliver everything for everyone should the need arise. However, as a result they often lack focus and end up doing nothing really well. It is important to realize that in order to succeed and perform well, a few and sometimes difficult trade-off decisions have to be taken. Organizations cannot deliver excellent performance if they attempt to be everything to everybody.

In this chapter, I will outline some tools that can be used to guide organizations through the analysis of the external environment to make decisions about outcomes and added values. These tools are individually well documented in other strategic management publications (which are referenced where appropriate). Individuals with a traditional strategy background will be familiar with these tools, and I have therefore tried to keep this section to the bare minimum. However, many public service and not-for-profit organizations embarking on a strategic performance management initiative might not be familiar with these tools. The application of these tools will provide the necessary information to understand the external organizational context and help clarify your stakeholder value proposition. Furthermore, these classical, externally focused tools are seldom brought together with the internal analysis (Chapter 3) to form a *cohesive* picture of an organization's strategic context.

IDENTIFYING KEY STAKEHOLDERS

An important component of strategy definition especially for any public service and not-for-profit organization is the identification of its key stakeholders
wherein you try to agree on who or what really matters for your organization.
I define a key stakeholder as a person, a group of people or an institution that
has an investment, share or interest in an organization and who may significantly influence the success of this organization. Public service and not-for-
profit organizations tend to have a large number of stakeholders; however, the
important aspect to look at here is whether or not they have a significant influence on the success of your organization.[2] Simple brainstorming sessions can
generate in excess of 50 stakeholders, but such a long list provides little guidance for the strategy definition. Therefore, it is important to narrow this list
down to the *key* stakeholders.

A knowledge about key stakeholders and their wants and needs will allow
organizations to shape their value proposition. In commercial enterprises,
shareholders have traditionally been the key stakeholders together with customers.[3] However, public service or not-for-profit organizations tend to lag this
shareholder focus, and instead, communities and governments tend to appear
on the top of the list. Possible key stakeholders for public service and not-for-
profit organizations could include:

- members of the public or a specific community,
- governments and government bodies,
- businesses and industry bodies,
- pressure groups, charities or NGOs.[4]

Each of these stakeholders may influence the success of your organization in different ways. Some might consume products or services, while others
might provide funding or resources. Typical roles of stakeholders include:

- customer or consumer of services and products,
- investor or fund provider,
- regulator,
- intermediary,
- distributor,
- employee and volunteer,
- supplier.

Some stakeholders might have a dual role. Government, for example, can
provide funding for an organization besides being a receiver of the services provided by the same organization. Three of my former colleagues have researched
the dual role of stakeholders in more detail and identified that most stakeholders
have a receiving as well as contributing function, that is they have something
they want from an organization (receiving function) and then there is something
an organization wants from them (contributing function).[5] Let's take customers

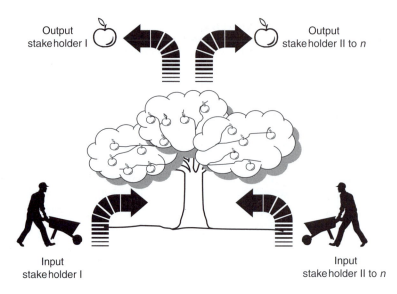

Output
stakeholder I

Output
stakeholder II to *n*

Input
stakeholder I

Input
stakeholder II to *n*

FIGURE 2.2   Output and input stakeholders.

for example; they might want high-quality services and products from an organization, while an organization wants loyalty and profitability from its customers. The same is true for employees. Employees might want job security and a decent salary, while the organization wants them to work hard and be innovative. In order to make this dual role more explicit, I suggest that you split your stakeholders into output stakeholders and input stakeholders.

Let me follow my tree analogy to explain what I mean by output and input stakeholders. Some stakeholders are those who eat the apples a tree produces, while other stakeholders are those who help to fertilize the tree and enable the tree to grow the apples in the first place. Figure 2.2 illustrates the distinction between output and input stakeholders. *Output stakeholders* are those who receive the benefits (outputs) of an organization's activity, for example consumers who receive services or products. *Input stakeholders* are those who contribute to the activities of an organization (provide inputs), for example employees, volunteers or suppliers who all have an input into the provision or production of services or products of an organization.

Some output and input stakeholders matter more than others, which is sometimes difficult to admit, especially in a political environment. To assess whether your stakeholders are in fact key persons or groups or not, I suggest you complete a simple analysis (see Fig. 2.3) that allows you to identify your organization's key stakeholders by splitting them into output and input stakeholders and by assessing their potential influence on the success of their organizations, together with a definition of what outputs they require or what inputs they provide.

Output Stakeholder	Output Required	Influence on Success (High, Medium, Low)
Customers		
Government		
Regulator		
Others		

Input Stakeholder	Input Provided	Influence on Success (High, Medium, Low)
Employees		
Partners		
Government		
Others		

FIGURE 2.3   Stakeholder analysis.

The stakeholder analysis can start with a brainstorming session to identify all major stakeholders. This list can then be split into output and input stakeholders. Please remember that it is likely that some of your stakeholders are both output and input stakeholders and therefore need to appear in each of the sections of your analysis table. Once the stakeholders are split, you can define what they require in terms of outputs or what they provide in terms of input. This should now allow you to assess their potential influence on the success of your organization. I suggest you use a simple scoring mechanism from high to medium to low influence on future success. I usually advise organizations I work with to cut the low-influence stakeholders and take the high-impact stakeholders as your key output and input stakeholders. As a rough guide, I would expect to see between four and six key stakeholders. If you have more than 10 or less than 3, then this is usually an indication of indecisiveness and a lack of strategic clarity. The medium-level stakeholders should also be ignored, but the list should be revisited from time to time to see whether any of them has moved up or down in the level of importance.

Both the key output stakeholders and key input stakeholders need to be taken into account when designing the forward strategy. The fact that they are important for the success of the organization warrants their place in the strategy and value creation model.

ASSESSING YOUR SECTOR AND ITS COMPETITIVE LANDSCAPE

Public service and not-for-profit firms operate in certain sectors in which they provide their services or products. It is important to analyze the competitive landscape of these sectors or industries. Such an assessment of the sector or industry is also referred to as microanalysis. The classic approach to analyzing the strategic microenvironment, that is how the organization relates to its industry or sector, was developed by Michael Porter in the late 1970s and early 1980s.[6] Since then, many organizations have applied his five forces framework (see Fig. 2.4), which Porter describes as a model for industry analysis, when examining their competitive environment.

While this model is appropriate for most not-for-profit organizations and public service providers that have a commercial aspect to their activities or compete with other organizations, it is not a particularly useful framework for analysis of government or public service organizations where the element of competition between rivals is absent. In fact, many government bodies or departments only exist because the service they provide would not be provided by anybody else such as commercial rivals.

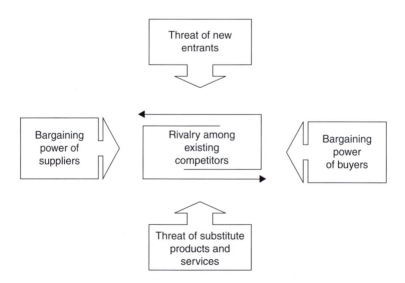

FIGURE 2.4   Porter's five forces framework.

Michael Porter's important contribution is to identify that competition is not only manifest in the industry's established combatants for market share, but also present in customers, suppliers, potential new entrants and substitute products. They are all influences that may be more or less prominent or active depending on the industry.

The likelihood of new entry looks at the extent to which barriers to entry exist. The more difficult it is for other organizations and potential competitors to enter a given market or sector, the more likely it is that existing organizations can enjoy the benefits and, for example, make relatively high profits. Many services provided by government departments, such as the national health service in some countries or the provision of rail services in others, would require heavy investments in infrastructure, which would make it unattractive for commercial rivals. Other elements such as a strong brand or a big market share in some sectors or industries make it unattractive for potential rivals because of the required marketing efforts and costs.

The power of buyers looks at the influence that customers have in an industry or sector. When buyer power is strong then they are more likely to force down prices and reduce the profits of firms that provide the product. Buyers are powerful if, for example, customers are concentrated, that is there are only few buyers each with significant market share. Also, when it is easy for customers to switch and use different products or services, the power that buyers hold increases.

The power of suppliers looks at the influence that suppliers have in an industry or sector. If suppliers are powerful, they can determine the terms and conditions on which business is to be conducted. Again, if the suppliers are concentrated, that is there are only few suppliers to choose from or there are high costs involved in changing suppliers, then this increases their influence and power.

The degree of rivalry looks at how much competition there is in one industry or sector. The more the rivalry in one area, the harder it is to generate high profits. Rivalry is high when many similar organizations compete with each other for the same customers, which often starts price wars.

The substitute threat looks at the availability of substitute products or services from outside the industry or sector. An often cited example is the fact that the aluminum beverage cans industry is constrained by the price and service competition from the glass bottles, steel cans and plastic container industry. It would be easy for any soft drinks producer to swap to either of these alternatives.

The five forces model therefore identifies that organizations are more likely to be successful in an industry or sector that is difficult to enter, there is limited rivalry, buyers are relatively weak, suppliers are relatively weak and there are few substitutes.

Learning what makes the business environment tick, therefore, is a vital piece of analysis. Many of the key determinants are illustrated in Table 2.1.[7]

TABLE 2.1 Porter's Elements of Industry Structure

Rivalry determinants	Entry barriers
Industry growth	Economies of scale
Fixed (or storage) costs/value added	Proprietary product differences
Intermittent overcapacity	Brand identity
Product differences	Switching costs
Brand identity	Resources requirements
Switching costs	Access to distribution
Concentration and balance	Absolute cost advantages
Informational complexity	Proprietary learning curve
Diversity of competitors	Access to necessary inputs
Corporate stakes	Proprietary low-cost product design
Exit barriers	Government policy
	Expected retaliation

Determinants of supplier power	Determinants of buyer power
Differentiation of inputs	Bargaining leverage
Switching costs of suppliers and firms in the industry	Buyer concentration versus firm concentration
Presence of substitute inputs	Buyer volume
Supplier concentration	Buyer switching costs relative to firm switching costs
Importance of volume to supplier	
Cost relative to total purchases in the industry	Buyer information
Impact of inputs on cost or differentiation	Ability to backward integrate
Threat of forward integration relative to threat of backward integration by firms in the industry	Substitute products
	Pull-through
	Price sensitivity
	Price/total purchases
	Product differences
	Brand identity
	Impact of quality/performance
	Buyer profits
	Decision makers' incentives

Determinants of substitution threat
Relative price performance of substitutes
Switching costs
Buyer propensity to substitute

ASSESSING THE ENVIRONMENTAL FORCES

In this section, we are trying to understand the external environmental forces that can affect an organization's operation but that are beyond its control. These include government regulations, the economy, demographics and social and cultural forces. Such an assessment of the external environment is also referred to as macroanalysis. A view of the strategic macroenvironment in which an organization operates is typically analyzed in a so-called PESTEL analysis that looks at the following factors: political, economic, social, technological, environmental and legal.[8] All of these factors can have a big influence on public service and not-for-profit organizations and the way they shape their strategy.

Political factors refer to government policies and include for example local or national funding policies, priorities about goods and services the government wants to provide, taxation policies, trade restrictions and tariffs, and investment incentives; the level of political stability may even be a significant issue in some markets. Political decisions can impact on many vital areas for organizations such as the education of the workforce, the health of the nation and the quality of the infrastructure of the economy such as the road and rail system.[8]

Economic factors, such as the rate of economic growth, interest rates, exchange rates and inflation rates, influence both an organization's cost of resources and its customers' demand for services and products. A rise in interest rates, for example, might increase the demand for social housing as less people would be able to pay their own mortgages, or it might squeeze the disposable income, which could affect the donation levels of charities.

Social factors include the demographic (e.g. population growth, age distribution, ethnic diversity, etc.) and the cultural aspects of the environments in which the organization operates. One example is the aging population in many Western countries. It is the fact that the life expectancy is constantly increasing and people living longer have an impact on, for example, health care systems.

Technological factors would typically include technological advances that have an impact on an organization and its levels of automation achievement and potential. With an ever-increasing rate of technological changes, there is a need for public service and not-for-profit organizations to keep up. While technology and automation can reduce costs and open the door for innovations, it can also change the demand for services and products. Just think of the demand for online services provided by government bodies.

Environmental factors include, for example, the weather, climate change or pollution levels. There is a growing desire to protect the environmental impact of organizational activities and many organizations are trying to reduce their carbon foot prints and water foot prints.

Legal factors include, for example, new legislations, regulations or disclosure requirements. Legal changes such as the introduction of age discrimination and disability discrimination legislation, an increase in the minimum wage and greater requirements for organizations to recycle are examples of relatively recent laws that affect an organization's actions. Legal changes can affect an organization's costs (e.g. if new systems and procedures have to be developed) and demands (e.g. if the law affects the likelihood of customers buying the good or using the service).[8]

Environmental and legal evaluations are frequently omitted from this type of assessment, but they are important contextual components. These factors

	Market 1	Market 2	Market 3
Political factors			
Economic factors			
Social factors			
Technological factors			
Environmental factors			
Legal factors			

FIGURE 2.5   PESTEL analysis.

can be assessed and evaluated for different markets or sectors (see Fig. 2.5). However, as Professor Andrew Gillespie points out:

> … it is important not to just list PESTEL factors because this does not in itself tell managers very much. What managers need to do is to think about which factors are most likely to change and which ones will have the greatest impact on them i.e. each [organization] must identify the key factors in their own environment. […] Managers must decide on the relative importance of various factors and one way of doing this is to rank or score the likelihood of a change occurring and also rate the impact if it did. The higher the likelihood of a change occurring and the greater the impact of any change, the more significant this factor will be to the [organization's] planning.[8]

This assessment of change in the future environment of the organization brings me to the tool of scenario planning, which I will discuss in the next section.

A LOOK INTO THE FUTURE

Another component that helps us to better understand the external environment is scenario analysis.[9] Scenario planning has been a recognized component of strategy formulation for a quarter of a century, since it became well known that companies like Shell had successfully applied it as part of their portfolio of strategic planning tools in the 1970s. Shell is generally accredited as the first company to use scenario analysis extensively for this purpose. However, Shell did not invent this approach; in fact, similar methods have been used for more than half a century. Herman Kahn developed the basic technique, which he initially called 'future-now' thinking, for the RAND Corporation in the 1950s.

At first, Shell applied scenarios to making better decisions about capital investment projects, which were more robust under a variety of alternative futures, before adapting them to strategic planning more generally. Shell of course is no stranger to the vagaries and impacts of volatile oil prices either then or today. Several practitioners from the company, including Pierre Wack, Kees van der Heijden and Arie de Geus, have described the art of applying scenario planning within organizations and the benefits that can be accrued. Shell's former managing director André Bénard commented in 1980 that 'Experience has taught us that the scenario technique is much more conducive to forcing people to think about the future than the forecasting techniques we formerly used'.

The basis of scenario planning involves defining and visualizing alternative views of how today's status quo in the operating environment might evolve in the future. It distills the countless possibilities of the future state into a limited set of coherent views. And of course, what might happen in the future has a corollary: what to do about it?

Typically, scenario analysis asks 'what if' questions about the future direction of the 'ecosystem' in which the enterprise operates. So both the macro- and microclimate analysis, described above, will normally be helpful toward informing this future-oriented analysis path. The power of scenario analysis, however, is in identifying the potential impacts of multiple events occurring simultaneously due to their interconnectivity. Each scenario needs a general theme that the organization's senior executives perceive as a potential threat – or opportunity – that needs to be addressed in the organization's longer-term strategic planning.

For example, Shell announced in June 2005 that the primary focus of its scenario planning to 2025 would switch from its previous (2001) assumptions about technological advances – that might see a shift in fuel consumption from oil to gas, nuclear and renewable energy – to one of national security and trust in the marketplace. Jeroen van der Veer, chief executive, said:

> Western societies now look to the state more than in recent decades to lead the restoration of physical security and market integrity. This brings into sharper focus the power of the state to regulate and coerce, in a role involving both the direct intervention to fight terrorism and police the market, and a more general emphasis on transparency, disclosure and good governance.

Like so many other fundamentally sound management techniques, this one is no different in that it can quite easily be abused. In the wrong hands, it can be used as a reason to procrastinate through overanalysis or, alternatively, the development of oversimplified scenarios can lead to underachievement in terms of its usefulness. Table 2.2[10] illustrates many of the best practice factors that should be taken into consideration for its proper application.

There are other limitations to scenario planning and these usually derive from a paucity of human imagination about what the future might hold. Keeping an open mind about future development and thinking 'out of the box' is critical for good scenarios. What happens when we apply our current models

---

TABLE 2.2 Seven Criteria for Good Scenarios

Decision-making power. Each scenario in the set, and the set as a whole, must provide insights useful for the question being considered. Most generic or general scenario sets lack this power and needed to be complemented for decision-making purposes.

Plausibility. The developed scenarios must fall within the limits of what future events that are realistically possible.

Alternatives. Each scenario should be at least to some extent probable, although it is not necessary to define the probabilities explicitly. The ideal is that the scenarios are all more or less equally probable so that the widest possible range of uncertainty is covered by the scenario set. If for instance only one of three or four scenarios is probable, you only have one scenario in reality.

Consistency. Each scenario must be internally consistent. Without internal consistency the scenarios will not be credible. The logic of the scenario is critical.

Differentiation. The scenarios should be structurally or qualitatively different. Thus, it is not enough for them to be different in terms of magnitude, and therefore only variations of a base scenario.

Memorability. The scenarios should be easy to remember and to differentiate, even after a presentation. Therefore, it is advisable to reduce the number to between three and five, although in theory we could remember and differentiate up to seven or eight scenarios. Vivid scenario names help.

Challenge. The final criterion is that scenarios really challenge the organization's received wisdom about the future.

---

and project these in the future is shown throughout history, where some of the biggest names in business have got the imagination of future scenarios totally and utterly wrong.[11]

In 1901, Daimler proclaimed, 'Worldwide demand for cars will never exceed one million, primarily because of a limitation in the number of available chauffeurs'. In 1915, Thomas Edison thought that fueled motors would soon be replaced by nickel–iron batteries. In 1945, Thomas Watson, then IBM's chief executive, declared, 'I think there is a world market for five computers'. In 1968, the respected *Business Week* magazine reported, 'The Japanese car industry isn't likely to carve a big slice out of the US market'. A decade later, the chief executive of Digital Equipment Corporation (DEC) was quoted as saying, 'There is no reason for any individual to have a computer at home'. And, as recently as 1995, Microsoft's Bill Gates famously commented, 'Internet is just a hype'. These misjudgments show that it is impossible to accurately predict the future. However, what we can use scenarios for is to identify possible and internally consistent developments in the external environment.

CLARIFYING THE STAKEHOLDER VALUE PROPOSITION

A stakeholder value proposition is a declaration of the way an organization proposes to use its resources and competencies to deliver a particular combination of values to its key output stakeholders. A value proposition brings together elements such as customer needs and organizational capabilities. This means that a definitive value proposition cannot be created until the internal analysis (see Chapter 3) is completed.

The traditional concept of value propositions was developed by Michael Treacy and Fred Wiersema in their book *The Disciplines of the Market Leaders* as a tool to differentiate the promises organizations make to their customers. The basic idea is that in order to be successful and focused in the delivery of services and products, organizations have to choose their main value proposition. The three customer value propositions organizations can choose from are operational excellence, product leadership and customer intimacy.[12]

Operational excellence means that organizations provide standard products to their customers, at the best price with least inconvenience. These organizations tend to offer the best prices for their products within their industry or sector. For example, Wal-Mart, Southwest Airlines and McDonalds would fall into this value proposition as they are good examples of operational excellence. This value proposition seems appropriate for many public service organizations and government bodies as they are trying to maintain good operational service levels in a world with ever-shrinking public funding.

Product leadership means providing the very best services and products (innovative service, new designs, new technology) to customers at the right time. Product leaders offer innovative products of exceptionally high quality, where price is not a significant barrier for their customers. Examples of companies with a product leadership value proposition include Sony and Apple. The product leadership value proposition can also apply to government and not-for-profit organizations. Examples might include the delivery of leading-edge care services provided by a cancer charity or the provision of the latest emergency response from fire departments and ambulance services.

Customer intimacy means providing the best total solution to customers. The organizations falling under this value proposition focus on delivering the best expert advice and tailored service to their customers,[11] for example McKinsey, Nordstrom Stores and IBM. Again, there are public service providers and not-for-profit organizations that would fall into this category as they provide customized advice in a close client relationship.

What is important to realize is that strategy is about choice and evaluating trade-offs. By choosing your value proposition, you help to clarify what your organization is offering to its customers. Again, public service and not-for-profit organizations sometimes find it difficult to make these choices.

The analysis of your output stakeholders (discussed earlier in this chapter) should help to understand what your customers want in terms of service and product delivery. The decision which of the three you choose sends a clear message to potential customers about the service levels and products they can expect. Once a value proposition has been chosen, it has an impact on other internal components such as your structure, core competencies, business process

TABLE 2.3 Value Propositions

Proposition		Strategic Objectives	Operational Objectives
Innovators	New innovative designs, products never seen before.	Provide breakthrough through generations of continuous new designs, new features within technological basis.	Long-term vision, robust R&D and product development, capacity to innovate within short product lifecycles.
Brand managers	Status from the product, they get lifestyle, a feeling of superiority.	Expand the market reinforcing the solid brand image of the product and the company.	Superb brand recognition. Focus market sector. Superior control over the product styles, quality and promotion.
Price minimizers	Ordinary, reliable products and services at lowest price possible. They get security on the product.	Production growth reaching high quality levels in the most cost-effective way and waste free.	Strong order fulfillment sustained by efficient and effective production processes within tight quality processes controls.
Simplifiers	Convenience and availability of the products. Hazard-free experience.	Build streamlined processes to make life simple and uncomplicated for customers in a novel and profitable way.	Strong availability. Superb order fulfillment – distribution by conventional and unconventional resources (networking, IT, etc.).
Technological integrators	Tailored products and services. They buy total solutions.	Tailor-specific and continuous solutions for carefully selected customers on the basis of permanent relationships.	Strong relationship with customer. Knowledge of customers' businesses, products and operations. Capacity to configure any specific need. Able to adopt the customer's strategy.
Socializers	Flexible services and inter-personal relationship because they trust in the company.	Build confidence and trust in the customers.	Sensitive fulfillment of customers' needs supported by careful delivery, reliability, and honesty. Excellent personal service.

and culture. Each of these are different for the various value propositions and this explains why it is not possible to deliver all of them at the same time.

Since its introduction many successful organizations have chosen their value propositions. However, choosing one value proposition does not mean ignoring the others. I suggest you choose one and concentrate on it but at the same time maintain acceptable levels of performance in the others.

Also, even though these three value propositions are widely accepted, there is often a need for hybrid models. There is nothing stopping you from designing your own unique value proposition. The danger, however, is that you are trying to do different things at the same time, which is always difficult.

One of my former colleagues, Dr Veronica Martinez, has extended the three traditional value propositions to build a value matrix.[12,13] The result of this is a more granular choice of six value propositions: innovators, brand managers, price minimizers, simplifiers, technological integrators and socializers (see Table 2.3). Reviewing these might help you decide what value proposition your organization delivers to its key output stakeholders.

SUMMARY

- In this chapter, I have discussed the importance of understanding the *external organizational context* to inform the forward strategy.
- Using the tools and classifications outlined in this chapter allows organizations to clarify the external environment and develop a better idea about their *outcomes* and *stakeholder value proposition*.
- An important starting point for public sector and not-for-profit organizations is the identification of their *key stakeholders*. Some stakeholders matter more than the others. Key stakeholders are those who have a significant impact on the future success of your organization.
- To enable a better analysis, I split stakeholders into *output stakeholders* and *input stakeholders*.
- *Output stakeholders* are those who receive the benefits (outputs) of an organization's activity, for example consumers who receive services or products.
- *Input stakeholders* are those who contribute to the activities of an organization (provide inputs), for example employees, volunteers or suppliers who all have an input into the provision or production of services or products of an organization.
- To understand the *microenvironment*, that is the competitive landscape of the industry or sector in which an organization operates, you can use Porter's *Five Forces* model to understand the likelihood of new entry, the power of buyers, the power of suppliers, the degree of rivalry and the substitute threat.
- To understand the *macroenvironment*, that is external environmental forces that are beyond the control of the organizations, you can apply a *PESTEL*

analysis, which allows you to assess potential changes in the political, economic, social, technological, environmental and legal contexts.

- The combination of the stakeholder analysis and an assessment of the macro- and microenvironment form a picture of the external context in which a stakeholder value proposition can be effectively devised, or, more commonly, revised.

- However, in order to finalize any stakeholder value proposition, it is important to ensure the organization is in a position to deliver on it, which will be the content of the next chapter.

REFERENCES AND ENDNOTES

1. See for example: de Bruijn, H. (2007). *Managing Performance in the Public Sector*, 2nd ed. Routledge, Oxford.
2. Stone, M., Semmens, A. and Woodcock, N. (2007). *Managing Stakeholders in the Public Sector*. CIPFA, London.
3. Rappaport, A. (1998). *Creating Shareholder Value: The New Standard for Business Performance*. Simon & Schuster, New York.
4. NGO stands for non-governmental organization and generally refers to not-for-profit organizations or networks within civil society that are proactive in the areas of development aid and development policy, and are particularly concerned to help establish (continuing) political and, wherever possible, financial independence.
5. See for example: Neely, A., Adams, C. and Kennerley, M. (2002). *The Performance Prism: The Scorecard for Measuring and Managing Business Success*. FT Prentice Hall, London.
6. See for example: Porter, M. E. (1980). *Competitive Strategy*. Free Press, New York; Porter, M. E. (1985). *Competitive Advantage: Creating and Sustaining Superior Performance*. Free Press, New York; Porter, M. E. (1979). How Competitive Forces Shape Strategy. *Harvard Business Review*, March–April, 137.
7. Source: Porter, M. E. (1985). *Competitive Advantage: Creating and Sustaining Superior Performance*. The Free Press, New York.
8. Gillespie, A. (2007). *Foundations of Economics – Additional Chapter on Business Strategy*. Oxford University Press, Oxford.
9. For an overview of how scenario planning can inform strategic performance management, please see: Fink, A., Marr, B., Siebe, A., Kuhle, J.-P. (2005). The Future Scorecard: Combining Internal and External Scenarios to Create Strategic Foresight. *Management Decision*, 43(2), 360–381.
10. Source: Lindgren, M. and Bandhold, H. (2003). *Scenario Planning*. Palgrave Macmillan, London, p. 31.
11. I borrowed these examples from my colleagues and friends Alexander Fink and Andreas Siebe, who have written a great book on scenario management (unfortunately only available in German): Fink, A., Schlake, O. and Siebe, A. (2001). *Erfolg Durch Szenario-Management*. Campus, Frankfurt.
12. Treacy, M. and Wiersema, F. (1996). *The Disciplines of the Market Leaders*. Harper Collins, London.
13. See for example: Martinez, V. (2003), "Understanding value creation: the Value Matrix and the Value Cube", PhD thesis, University of Strathclyde, Glasgow.

Chapter 3

Understanding Inputs, Competencies and Resources

The third component of the contextual analysis requires looking inside the organization in order to make a critical appraisal of its competencies and key resources. Compared to the external analysis discussed in the previous chapter, this internal part of the strategic analysis is relatively new and many organizations are struggling with the identification of their competencies and resources that enable them to perform well. In particular, many public service and not-for-profit organizations seem to have difficulties with the identification of their intangible enablers. There is immense confusion among managers about what intangibles are and how to classify them, as well as the difference between competencies, capabilities and resources. I therefore aim to provide a detailed discussion and breakdown of organizational resources, and how they form the foundation for capabilities and competencies, before moving on to look at the tools to understand and map these.

Staying with the tree analogy, this chapter therefore deals with the roots (the resources) and the trunk (the competencies). The questions we are trying to address include (see Fig. 3.1):

- What are we good at?
- What are our competencies and capabilities?
- What kind of services and products is our organization capable of producing?
- What does our resource architecture (tangible and intangible) look like?
- How do resources combine to give us our capabilities?

 =

FIGURE 3.1 Understanding the internal context.

YOUR ORGANIZATIONAL RESOURCES

We start with the foundation and look at the roots of the tree – which represent the organizational resource architecture that enables an organization to perform. Resources are critical building blocks of strategy because they determine not what an organization wants to do but what it can do.[1] Even though economists started to make a strong case for the significance of intangible resources as an important production factor in the early part of the nineteenth century,[2] organizations have traditionally looked at only their financial and physical resources and, by doing so, often overlooked their intangible resources as an enabling force and a source of competitive performance. Today, most executives see the critical importance of intangible resources as the drivers of performance. Based on this, we can classify organizational resources into the following three principal categories (see Fig. 3.2): financial resources, physical resources and intangible resources.

Financial resources are simply the amount of funding or finance available. While some organizations in the government or not-for-profit sector can rely only on their allocated budget, others are able to generate further monetary resources from various sources such as services and product sales, donations, borrowings, asset sales and equity stakes.

Physical resources consist of items such as buildings, information and communication technology infrastructure, plant and equipment, premises or land and, in some cases, owned natural resources. Examples might include pet hospitals in a pet aid charity such as the PDSA or the natural resources owned and maintained by the National Trust in the United Kingdom.

Intangible resources are nonphysical sources of value such as knowledge and skills of employees, brand image, reputation, relationship with suppliers, organizational culture, best practices or intellectual property. These are becoming increasingly important for public service and not-for-profit organizations.

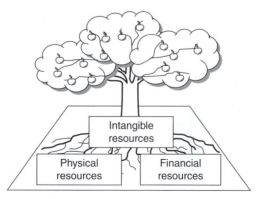

FIGURE 3.2   Organizational resources.

From a resource-based perspective, it is argued that resources are valuable only if they provide a unique competitive advantage for the organization and if they support the organization's core competencies.[3] In order for them to be strategically valuable, resources must therefore be inimitable, not substitutable, tacit in nature or synergistic.[4] It is often argued that in today's economy, most physical resources are transient and are therefore rarely sources of competitive advantage. With this in mind, one might argue that they could therefore be ignored. However, many intangible resources are valuable only in relationship with existing physical resources, or vice versa. Physical resources are often levers that enable companies to benefit from their intangible resources.

Wal-Mart's expertise in inventory replenishment, for example, is a key intangible element, but without physical resources such as the innovative physical distribution centers and store layouts, it would be less, or not at all, valuable.[5] The fact that physical resources rarely create a competitive advantage on their own does not mean that they cannot be key drivers of competitive advantage and performance. One might think about DeBeers and its possession of diamond mines or oil companies such as Exxon or BP and their oil reserves. We will come back to the interrelated nature of organizational resources, but before we do this, we need to better define intangible resources and their important role in creating value.

THE IMPORTANCE OF INTANGIBLE RESOURCES

Intangibles are important enablers of performance and success. A recent survey commissioned by the consulting firm Accenture revealed that most executives around the world believe that intangibles are critical for the future success of their businesses.[6] However, at the same time, most agreed that their approaches to managing intangibles were poor or nonexistent. Indeed, it is estimated that the level of US corporate investment in intangible assets, around $1 trillion annually, almost matches that of investment in tangible assets.[7]

Not only commercial enterprises are seeing the value in intangible resources, but also other organizations and governments are recognizing the importance of them. In the United Kingdom, for example, the UK prime minister wrote in a recent government white paper that intangible resources such as creativity and inventiveness are the greatest source of economic success but too many firms have failed to put enough emphasis on R&D and developing skills.[8] The UK secretary of state for trade and industry added in a recent report that, increasingly, it is the intangible factors that underpin innovation and the best-performing businesses.[9] A report of the Brookings Task Force on Intangibles outlines that the large and growing discrepancy between the importance of intangible assets to economic growth and the ability to clearly identify, measure and account for those assets is a serious problem for business managers, investors and governments.[10] An important public sector study of about 100 local authorities in Israel finds that intangibles such as managerial capabilities, human capital, internal

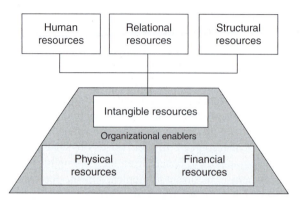

FIGURE 3.3   Classification of intangible resources.

auditing, labor relations, organizational culture and perceived organizational reputation are actually important enablers of good performance.[11] In order to identify intangibles, we need to first define what we mean by them.

WHAT ARE INTANGIBLE RESOURCES?

The concept of intangible assets is frequently used but not always well defined. Often different terms are used to describe the same concept, which means that intangible resources are also referred to using terminology such as 'intangible assets', 'intellectual capital' or 'knowledge assets'.[12] Since this book is about strategic performance management, we use the terminology 'intangible resource', which is most closely associated with strategic management thinking.[13]

Many different classifications and definitions exist for intangible resources. It is important to stress that there is no right or wrong classification. Instead, what is important is that a classification is comprehensive and doesn't leave out important forms of intangible resources. The classification provided below ensures that all critical intangible resources are included. The key objective of this classification is to facilitate the identification of the intangible resources within organizations. Debates on whether one intangible should be put into one category or another are therefore not productive or particularly useful. What is important is that we identify all intangible resources that matter.

Here, intangible resources are defined as non-tangible resources that are attributed to an organization and that support an organization's competencies and therefore contribute to the delivery of the organizational value proposition to its various stakeholders. Intangible resources can be split into three component classes: human resources, structural resources and relational resources (see Fig. 3.3). Table 3.1 provides examples of different intangible resources under each of the three categories.

TABLE 3.1 Examples of Intangible Resources

Human resources	Relational resources	Structural resources
Knowledge and skills	Formal relationships	Organizational culture
Education	Informal relationships	Corporate values
Vocational qualification	Social networks	Social capital
Work-related experience	Partnerships	Management philosophy
Work-related competencies	Alliances	Intellectual property
Emotional intelligence	Brand image	Brand names
Entrepreneurial spirit	Trust	Data and information
Flexibility and changeability	Corporate reputation	Codified knowledge
Employee loyalty	Customer loyalty	Patents/copyrights
Employee satisfaction	Customer engagement	Trade secrets
	Licensing agreements	Processes and routines
	Distribution agreements	Formal processes
	Joint ventures	Tacit/informal routines
		Management processes

HUMAN RESOURCES: SKILLS, KNOWLEDGE AND STAFF ENGAGEMENT

The principal subcomponents of an organization's human resources are naturally its workforce's skill sets, depth of expertise and breadth of experience. Human resources can be thought of as the living and thinking part of the intangible resources.[14] These resources therefore walk out at night when people leave, whereas relational and structural resources usually remain with the organization even after people have left. Human resources include the skills and knowledge of employees, as well as know-how in certain fields that are important to the success of the enterprise, plus the aptitudes and attitudes of its staff.[15] Employee loyalty, motivation and flexibility will often be a significant factor too since an organization's 'expertise and experience pool' is developed over periods of time; for example, high levels of staff turnover mean that an organization is hemorrhaging these important resources. The importance of knowledgeable and experienced staff has been demonstrated by many studies. For example, local authorities implementing strategic human resource management practices that result in organization-specific educated and trained employees have been found to outperform those that do not implement such practices.[16]

RELATIONAL RESOURCES: PARTNERSHIPS, IMAGE AND CORPORATE REPUTATION

Relational resources are the relationships that exist between an organization and any outside party, both with key individuals and other organizations. These can include customers, intermediaries, employees, suppliers, alliance partners, regulators, pressure groups, communities, creditors or investors. Relationships tend to fall into two categories – those that are formalized through, for example, contractual obligations with major customers and partners and those that are more informal. While, in the past, the former tended to be predominant, today the latter have a more important say in how the enterprise is managed. The type of relationship can have an impact on the value of these relationships; for example, they can determine the effectiveness of the information that is transferred between related parties.

Research has confirmed that, for example,

> Good labor relations between the management and the employees are likely to improve the performance of a local authority [...] upholding principles such as fairness, safety, and trust is likely to motivate members to higher production and overall standards, thereby countervailing the effects of inefficiency, ineffectiveness, and concealed unemployment that too often burden local authorities ....[11]

Other factors that fall into this category are brand image, corporate reputation and product/service reputation. Increasingly, this latter subcategory can be particularly important to the success or failure of government and not-for-profit organizations. In fact, a recent study found that perceived organizational reputation is a key success factor in local authorities:

> Highly reputable local authorities can, for example, attract new residents from the higher socioeconomic levels, as well as new investors, and thus strengthen their fiscal state, create new jobs, and offer a higher standard of living.[11]

Reputation is important for most public service and not-for-profit organizations. Hospitals rely on a positive corporate reputation in the same way schools, police forces and fire departments do. Charities rely heavily on corporate reputation and a dip in reputation can seriously impact donation levels.

STRUCTURAL RESOURCES: INFORMATION, CULTURE AND PRACTICES

An organization's structural resources cover a broad range of vital factors. Foremost among these factors are usually the essential operating processes, the way it is structured, its policies, its data and information and the content of its databases, its leadership and management style and its culture, as well as intangible resources that are legally protected by patents or copyrights, for example. The structural resources can be subcategorized into organizational culture, practices and routines and intellectual property.

Organizational culture can reinforce the achievement of the overall goals, sometimes also referred to as social capital and context.[17] Government as well as not-for-profit organizations often suffer from a bureaucratic or hierarchical organizational culture and could benefit from a greater emphasis on change, flexibility, entrepreneurialism, outcomes, efficiency and productivity.[18] The right corporate culture gives each person in an organization a common and distinctive method for transmitting and processing information; it defines a common way of seeing things, sets the decision-making pattern and establishes the value system.[19] Culture resources embrace categories such as corporate culture, organizational values and management philosophies. They provide employees with a shared framework to interpret events, a framework that encourages individuals to operate both as an autonomous entity and as a team in order to achieve the company's objectives.[20] For example, research shows that a government or not-for-profit organization with a strong organizational culture that emphasizes elements such as high involvement of the organization's members, shared beliefs, ability to adapt to the environment and a sense of mission is likely to perform better than its counterparts lacking such an organizational culture.[21]

Practices and *routines* can be important organizational resources. Shared knowledge in organizations is expressed in routines and practices.[22] Practices and routines include internal practices, virtual networks and review processes; these can be formal or informal procedures and tacit rules. Formalized routines include process manuals providing codified procedures and rules; informal routines could be codes of behavior or understood (but unstated) workflows. Practices and routines determine how processes are being handled and how work flows through the organization. An example of a process that has become a valuable strategic resource is the 25-min airplane turnaround time at Southwest Airlines. A process introduced as a necessity to start up the business as a low-cost carrier, today has become a key differentiator.

Intellectual property – owned or legally protected intangible resources – is becoming increasingly important. Here, intellectual property is defined as the sum of resources such as patents, copyrights, trademarks, brands, registered designs, trade secrets, database contents and processes whose ownership is granted to the company by law.[23] Intellectual property is an element of organizational knowledge that is owned by the organization and can't walk out at night when everyone goes home. It represents the tools and enablers that help to define and differentiate an organization's services and product offering. Intellectual property includes trademark symbols such as the McDonald arches and the Nike swoosh or the patented '1-click' buying option at Amazon.com. Coca-Cola, for example, made a conscious decision to keep the formula for Coke a trade secret that is actively protected. Had they patented the formula instead, their patent protection would have run out many years ago, most likely destroying their market share. Many government and not-for-profit organizations possess intellectual property in the form of information and database contents, brand names and trademarks.

Even though most organizations possess a wide stock of intangible resources, not all of those are critical value drivers and enablers of successful performance. The reasons for this are that the value of resources is context specific and resources are not just static – they dynamically interact with each other to be transformed into capabilities and core competencies. The latest knowledge of how to treat patients with heart conditions is critical for hospitals and health services, for example, but the same knowledge is of little value to a court of justice or a nature reserve trust (context-specific nature of resources). Having the knowledge of how to treat patients with heart conditions is valuable to a hospital only if it also has surgeons with the skills to apply this knowledge and the physical infrastructure needed to treat patients. This is often referred to as the interconnectedness of organizational resources.

CAPABILITIES AND CORE COMPETENCIES

No discussion on strategy and organizational resources would be complete without a view of how the individual resources interrelate with each other to create vital capabilities and core competencies. Similar to the definition of intangible resources, little consensus exists about what exactly constitutes a capability or a core competence. While people often use the words 'capabilities' and 'core competencies' interchangeably, I believe that some clear distinctions need to be made, which I outline below.

Capability refers to the quality of being capable, physically or intellectually. Capabilities are localized bundles of brilliance – sometimes called 'centers of excellence' – that shine within their organizations through a combination of the resources. They are activities that an organization is able to perform better than other activities. Capabilities can be defined as the combination of a set of organizational resources (physical, monetary, human, relational and structural) that collectively enable that organization to perform well in specific areas. A number of combinations of different organizational resources can therefore yield many different capabilities that the organization is good at but that may or may not be strategically vital.

Core competence, on the other hand, is an excellently performed internal activity that is central, not peripheral, to a company's strategy, competitiveness and value proposition. The difference between capabilities and core competencies is that an organization might have many potential capabilities resulting from the combination of their resources, whereas it will have very few core competencies. Core competence is therefore a capability, or a set of capabilities, that is linked to the strategic value proposition of an organization (see Fig. 3.4).

The term 'core competencies' first became prevalent following an award-winning article published in the *Harvard Business Review* by Hamel and Prahalad.[24] Core competencies (sometimes referred to as 'strategic core competencies') are therefore a distinctive combination of organizational resources, such as applied technologies, skill sets and/or business processes, which have evolved and been learned over a period of time in response to customer and other key stakeholder needs.[25] They uniquely define an organization and provide a thread running through it, weaving their resources together into a coherent whole.[26] It is also the evolutionary learning process that gestates within the organization that makes it unique and, therefore, extremely difficult for others to replicate.

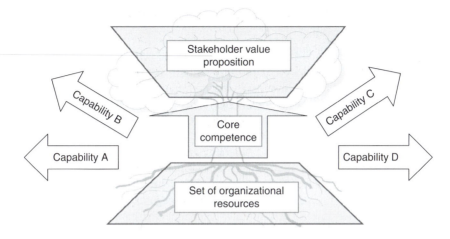

FIGURE 3.4 Capabilities versus core competence.

Often-quoted examples of core competencies are Sony's expertise in miniaturization to create 'pocketable' consumer products and Honda's ability to get the best out of internal combustion engines.[27] Honda's superior expertise in compact engine construction enabled it to build on this competence and become a major competitor to many established manufacturers of, for example, motorboat engines, lawn mowers and portable generators through its retail relationships, while continuing to sell automobiles and motorcycles through its own dealerships. Another company's management that has built its success on its understanding of the firm's core competencies is Wal-Mart, which transformed the company by building on its knowledge of 'cross-docking'. This supply chain management best practice – whereby incoming goods are immediately trans-shipped to delivery vehicles without expensive warehousing – was one of the factors that enabled Wal-Mart to outperform its rival K-Mart.[5]

Recognizing what capabilities an organization possesses and what core competencies are required to deliver the organizational value proposition is fundamental to strategy development. What can pose a challenge is harmonizing different capabilities that are required to deliver desired outcomes or core competencies, for example, excellent customer service. Consider a common business process, such as the order-to-cash fulfillment process in a public library for instance. The customer places an order, and the library orders or locates the publication, delivers it and then gets paid for it. To the customer, it is a single process with multiple components, but it implies the presence of at least six different capabilities:

- Customer order handling capability
- Planning and scheduling capability
- Procurement capability

- Location capability
- Distribution capability
- Credit management capability

Each of these capabilities requires different skill sets, practices, technologies (although some IT systems will likely be multifunctional and integrated) and physical resources, such as offices and the library. If each component capability does not dovetail effectively and cohesively with its peers within the same business process, then the customers will not be satisfied. The same applies to other similar processes, such as demand generation, product/service development and after-sales service. However, effectively joined-up capabilities can be an incredibly powerful means of creating a distinctive competitive advantage. Therefore, capabilities not only work closely together to deliver immediate process outputs for an enterprise's stakeholders but also, consequently, have longer-term outcomes that can dictate the enterprise's strategic options.

Hamel and Prahalad recommend three tests to help identify core competencies. The core competencies must:

- provide potential access to a wide range of markets,
- make a significant contribution to the perceived customer benefits of the end product (or service),
- be difficult for competitors to imitate.

It is important to recognize too that core competencies often do not reside in a single strategic business unit and, therefore, corporate offices need to take ownership for them to ensure that they are properly nurtured so that they do not inadvertently wither or even get outsourced. The notion of strategic core competencies requires that, over time, some existing core competencies will need to be abandoned, others strengthened and some new ones created. This is done by reconfiguring the underlying resource architecture that shapes the core competencies – this could mean, for example, retraining people, updating the IT infrastructure, redesigning processes, establishing a new reputation or building new strategic alliances.

THE DYNAMIC NATURE OF RESOURCES

The above outline of organizational resources addresses the stock of resources and helps organizations to identify what key resources they have. This discussion highlighted that resources need to be bundled together to form either capabilities or competencies. Therefore, in order to be valuable, organizational resources have to be transformed, through core competencies, into products or services that deliver value. Resources are often referred to as performance drivers or enablers, which reinforces the notion of causal relationships between the resources and organizational value creation. Intangible resources such as employee skills and customer relationships often deliver customer satisfaction and loyalty, which in turn deliver successful organizational performance levels.[28]

Individual resources often impact performance with 'causal ambiguity'.[29] This means it is difficult to identify how individual resources – let's say the corporate reputation – contribute to success without taking into account the interdependencies with other resources. For example, the value of corporate reputation might depend on the quality of the service delivery process and the service reputation itself. The latest technology can be an important resource in organizations, but it is worth little without the right knowledge and competencies of how to operate it. In turn, all the latest understanding and knowledge of how to operate technology is worthless if employees do not have access to the technology.

Baruch Lev, professor at New York's Stern School of Business, notes that intangibles are frequently embedded in physical resources (e.g. the technology and knowledge contained in an airplane) and in labor (e.g. the tacit knowledge of employees). This leads to considerable interactions between tangible and intangible resources in the creation of value. He also emphasized that 'when such interactions are intense, the valuation of intangibles on a stand-alone basis becomes impossible'.[30] This is why a balance sheet approach to organizational resources does not provide information on the important interrelationships between them.[31] To gain strategic insights into the importance of organizational resources, it is important to understand their interdependencies with other resources to form core competencies and thus products and services that deliver value.

In conclusion, this means that organizational resources, both tangible and intangible, interact with and depend on each other to form the basis for capabilities and core competencies. Organizations, therefore, require tools to help them understand their resource architecture, capabilities and core competencies.

ASSESSING YOUR STRENGTHS AND WEAKNESSES

One tool that has long been used by public sector and not-for-profit organizations to understand the strengths and weaknesses of their organizations is the SWOT analysis – SWOT stands for strengths, weaknesses, opportunities and threats.[32] Opportunities and threats are external factors and should be derived from the external analysis (Chapter 2). Here, the focus is on the strengths and the weaknesses since they can provide input for a contextual understanding of the internal environment. This analysis can play an important role in bringing together a first overview and consensus opinion of what really matters and what the organization is good at or not so good at. It can help to bridge the factions (e.g. technical vs. marketing vs. finance) that can exist within most large organizations. Figure 3.5 illustrates several of the typical components that might be brought together to summarize the internal and external views of the organization for strategic decision-making purposes. The content is intended to be illustrative rather than authoritative and other considerations may come into play.

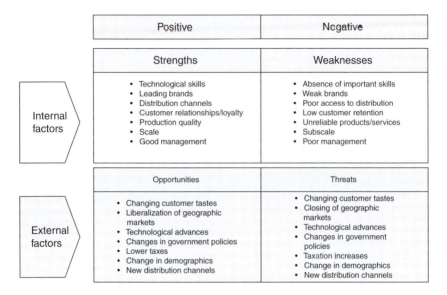

	Positive	Negative
	Strengths	**Weaknesses**
Internal factors	• Technological skills • Leading brands • Distribution channels • Customer relationships/loyalty • Production quality • Scale • Good management	• Absence of important skills • Weak brands • Poor access to distribution • Low customer retention • Unreliable products/services • Subscale • Poor management
	Opportunities	Threats
External factors	• Changing customer tastes • Liberalization of geographic markets • Technological advances • Changes in government policies • Lower taxes • Change in demographics • New distribution channels	• Changing customer tastes • Closing of geographic markets • Technological advances • Changes in government policies • Taxation increases • Change in demographics • New distribution channels

FIGURE 3.5   Identifying strengths and weaknesses.

ASSESSING WHAT RESOURCES YOU POSSESS IN YOUR ORGANIZATION

The above categorization of resources that I have outlined can be used to facilitate a discussion about the current stock of resources. This can be conducted in individual interviews with key people in the organization; it can also be done in facilitated workshops, or even via a mail or online survey. From experience, doing individual interviews or surveys works best, as it means that all participants have their say, without their opinion being suppressed by stronger or more dominant participants in workshops. The resource categorization can be used to create a template that guides people through the different resources and prompts them to think about the different types of resources in their organization (see Fig. 3.6).

It is important to emphasize again that the objective of this resource classification template is to address as many different resources as possible and facilitate a discussion. Therefore, it doesn't matter whether we classify a resource as relational, structural or human. The aim here is to stimulate awareness of possible resources in order to identify the resource stock of an organization, rather than to put them into rigid categories.[33] For the purpose of identifying resources, it therefore isn't important if, for example, we put 'the ability to build customer relationships' into the human resources or the relational resources category. What matters is that we create a realistic picture of the existing resource architecture.

The individual responses can then be brought together and a list of all the resources can be presented in a facilitated workshop. Participants in this workshop are usually the senior managers; however, sometimes it is also advisable

Resource category:	Sub categories:	Resources with a significant presence in our organization:
Physical resources	Property, plants, location of buildings, information and communication infrastructure, machines, equipment, natural resources, physical infrastructure, office design, etc.	
Human resources	Education, technical knowledge and expertise, skills, know-how, attitudes, experience, motivation, flexibility, commitment, creativity, etc.	
Relational resources	Customer relationships, supplier relationships, reputation, image, trust, contractual relationships, informal relationships, alliances, relationships with regulators, partners, etc.	
Structural resources	Processes, tacit routines, organizational structure, governance and management approaches, organizational culture, social capital, shared identity, patents, brand names, copyrights, trade secrets, codified information and knowledge, e.g. in databases or process manuals, etc.	
Financial resources	Cash, investments, bonds, loans, budget, etc.	

FIGURE 3.6   Identifying your resource stock.

to include people from different hierarchical levels or even external stakeholders. This might depend on the complexity of your organization. The more complex the business is, the better it is to involve as many people as possible. The outcome from this exercise is a list of key resources. At this point, it is not important anymore to use the categories, but rather the individual resources, presented in a language that is understood within the particular organization. Different organizations tend to have different names, or organization-specific terminology, to describe the same resources. It is always advisable to use the language that is used within the organization instead of the categories or examples provided in the template above. Using terminology such as 'human capital', for example, can cause misunderstanding or even cynicism, especially if this terminology is not usually used within the organization.

IDENTIFYING THE RELATIVE IMPORTANCE OF KEY RESOURCES

The relative strengths or importance of the identified key resources can be assessed only in the context of the existing strategy. The questions to answer are: How important are our different existing resources to achieving our over-all value proposition? How strong are our existing resources and how can we utilize them more effectively? Here is where the market- and resource-based views come together from opposite sides. The former starts with the strategic value proposition identified by the external analysis (Chapter 2) and then identi-fies the relative importance of each resource to achieve the strategic objectives. The latter looks at the existing resource architecture and evaluates the strength of each resource in the organization independently of any value proposition or existing opportunities in the market.

The most realistic situation in government agencies and most not-for-profit organizations is that they have a more or less prescribed list of products and serv-ices and tend to have relatively narrow remits. There will only rarely be a situ-ation in which organizations discover that they possess valuable resources that would allow them to enter completely different sectors or industries. If you are in a situation where your value proposition is prescribed, or where you feel your existing strategy is good or can't be changed, you start the assessment of your resources with your current value proposition in mind. If you start with a blank piece of paper, which I have to admit I have never seen, you could then evaluate the strength of your resources without any context. In all other cases, you might want to do both.

Assessing the importance of the different resources to deliver your value proposition and assess your resource strengths independently allows organiza-tions to perform a gap analysis. This lets you understand whether you are build-ing the appropriate resource architecture for your value proposition, or whether you are under- or over-investing in certain areas (for a more in-depth discussion on this, see risk assessment in Chapter 6).

This assessment is best done individually, either in interviews or by survey, or it can be completed as part of a workshop. The easiest way to do this is to pro-duce a list with the resources identified above and then to add columns for people to assess the importance (see Fig. 3.7). Here, both assessments are included, which allows a gap analysis.

The results from the individual assessments can then be aggregated and visualized in a resource map. Such a map is the visual representation of the relative strength or importance of the different resources. It is also possible to include the two data sets (strengths and importance) and visualize the different size bubbles to indicate any gaps.

Figure 3.8 shows an illustrative commercial example of such a map that was created for a leading online-retailing business.[34] Its aim was to understand the relative importance of its resources to deliver its existing value proposition. The value proposition for this well-known retailer was to become the world's pre-ferred source for a particular type of goods by providing consumers not only with top-level service but also with high quality of value-added information, excellent price, simple transactions and an enjoyable shopping experience. In this exam-ple, structural and human resources were the most important value drivers for this

Identified key resources	Relative strengths of these resources in our organization 0 = not at all important 10 = vitally important	Relative importance of these resources to delivering our value proposition 0 = not at all important 10 = vitally important
Our specific subject knowledge	7	10
Our perceived reputation	4	9
Relationships with key partners	4	6
Our buildings	9	2
Our website	8	7

FIGURE 3.7   Assessing the importance of resources.

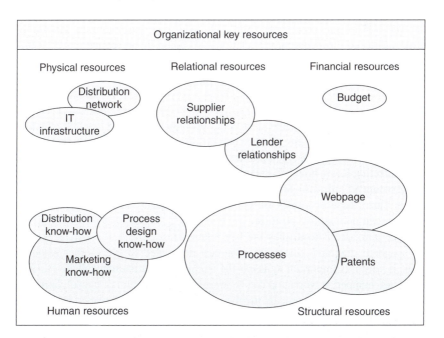

FIGURE 3.8   Visualizing the relative importance of the key resources.

commercial enterprise, with particular emphasis on its know-how of the market and its customers, plus its processes and Web site. Other important resources were relational, especially with goods suppliers and moneylenders, as the business is still in the growing phase and not making any profits. This map helped the organization to make sure that it is appropriately allocating its resources.

ASSESSING THE INTERDEPENDENCIES OF YOUR RESOURCES

Resources depend on and interact with each other in order to create a core competence, which in turn helps to deliver the value proposition. This means that resource interdependencies can be assessed only in relation to the existing core competencies and value proposition of the organization. If you have defined your core competence, you can then use the resource list created above to understand how the resources interlink to deliver core competencies and value proposition.[35]

A study of government organizations clearly finds that organizational resources (especially intangibles) enhance each other in their effect on organizational performance, which suggests that organizations seeking to maximize their performance subject to a given budget should balance the development of their organizational resources. As a consequence, organizations should not only attempt to estimate the direct effect of each organizational resource on performance but also assess the effects on performance of the interactions among them.[11]

Individuals can use a matrix to rate how resource A is dependent on resource B to deliver the core competence, until all resource combinations are assessed.[36] The scale used for assessing the relationships could be between 0 and 5, with 0 indicating no relationship and 5 indicating a very strong dependency. Again, these matrices can be completed by individuals and then aggregated.

The results allow a facilitator to create a resource map with interdependencies between resources (see Fig. 3.9). Here, thinner arrows indicate a weaker dependence and thicker arrows a stronger dependence. In this example, the processes, so vital for the core competence of this commercial business (e.g. provision of a simple and enjoyable online shopping experience), strongly depend on the marketing know-how and understanding of the customer needs, as well as on the existing IT infrastructure, the distribution know-how and supplier relationships. Once these processes are created, this company aims to patent them to protect their competitive position.

The aim here is to display the key relationships. Often, these maps start as a 'spaghetti diagram' where all resources are related to every other resource. This is why it is important to assess the strength of the interdependencies. Once these are mapped, a workshop can be arranged with the aim of consolidating the interdependencies. The process involves jointly identifying the key dependencies and eliminating minor and unimportant dependencies. The objective is to gain consensus on the final map to represent the key resources and their key interdependencies that together deliver the core competence.

ASSESSING THE FUTURE OF YOUR RESOURCES

As discussed in Chapter 2, scenarios are traditionally used to describe possible alternative future developments in the external environment, which then inform the positioning and stakeholder value proposition. However, with

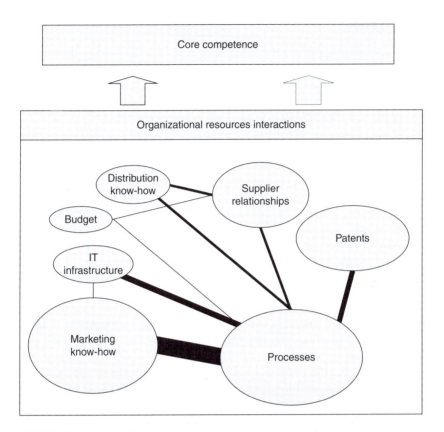

FIGURE 3.9   Interdependence of resources.

a shift in focus toward a resource-based view of strategy, scenarios can also be used to describe alternative internal development paths for an organization.[37] Internally, organizations have many options of how to react to external opportunities and threats. As outlined above, competencies are based on the resources that often depend on each other and are often path dependent – that is, present choices about options are influenced by past choices.[38]

Within an organization, there are often different perspectives on current problems and unsolved conflicts, different assumptions about the levers where changes could be initiated, different prioritizations of capability and resource development and different interests. Considering these different views and accumulating diverse accounts about the possible developments of the internal resources can yield important information about the future direction of the organization. Holding a workshop to explore future developments of the organization allows managers to bring in their personal ideas and visions of the organization's future and systematically link them to several strategy scenarios. Such workshops are best led by an external facilitator who is experienced in developing scenarios.

SUMMARY

- This chapter has emphasized the importance of understanding the internal resource architecture and core competencies of an organization.
- Using the tools and classifications outlined in this chapter allows organizations to clarify the internal organization context and develop a better understanding of their *resource architecture* and *core competencies*.
- *Resources* are enablers and drivers that underpin future performance. They can be categorized into *monetary*, *physical* and *intangible* resources.
- In order to facilitate the identification of your resource stock, *intangible resources* are split into *human*, *relational* and *structural* resources.
- *Human resources* relate to the people in the organization and include knowledge, skills, employee loyalty, staff engagement and experience.
- *Relational resources* relate to relationships between the organization and external parties. These include partnerships, licensing agreements and corporate reputation.
- *Structural resources* relate to internal practices and processes as well as to intellectual property and organizational culture. This category includes copyrights, data and information and leadership style.
- *Organizational resources are interdependent*. All key resources interact with and depend on each other to form the basis for capabilities and core competencies.
- Core competencies underpin the delivery of the stakeholder value proposition.
- Core competencies differ from capabilities. While a *capability* refers to any excellently performed internal activity the organization is capable of, a *core competence*, on the other hand, is an excellently performed internal activity that is central, not peripheral, to a company's strategy, competitiveness and value proposition.
- The internal analysis, together with the analyses of the boundary conditions (Chapter 1) and the external environment (Chapter 2) form the basis for the strategy. In the following chapter, I will discuss how we translate the output from the first three chapters into an integrated and coherent organizational strategy.

REFERENCES AND ENDNOTES

1. Collis, D. J. and Montgomery, C. A. (1997). *Corporate Strategy – Resources and the Scope of the Firm*. McGraw-Hill, Boston, p. 9.
2. See for example: Senior, N. W. (1836). *An Outline of the Science of Political Economy*. Longman, London; Marshall, A. (1890). *Principles of Economics*. Macmillan, London. Volume I (1982): Knowledge and Knowledge Production; Volume II (1982): The Branches of Learning; Volume III (1984): The Economics of Information and Human Capital (Posthumous).
3. See for example: Wernerfelt, B. (1984). A Resource-Based View of the Firm. *Strategic Management Journal*, 5(2), 171.

4. See: Barney, J. B. (1991). Firm Resources and Sustained Competitive Advantage. *Journal of Management*, 17(1), 99; Rumelt, R. P. (1984). Towards a Strategic Theory of the Firm. In: *Competitive Strategic Management* (Lamp, R. B., ed.), pp. 89–102. Prentice Hall: New Jersey; de Haas, M. and Kleingeld, A. (1999). Multilevel Design of Performance Measurement Systems: Enhancing Strategic Dialogue throughout the Organization. *Management Accounting Research*, 10, 233–61

5. Stalk, G., Evans, P. and Shulman, L. E. (1992). Competing on Capabilities: The New Rules of Corporate Strategy. *Harvard Business Review*, 70, 57–69.

6. Molnar, M. J. (2004). Executive Views on Intangible Assets: Insights from the Accenture/ Economist Intelligence Unit Survey. Accenture Research Note *Intangible Assets and Future Value*, Issue 1, April.

7. Lev, B. (2002). Intangibles at a Crossroads: What's Next? *Financial Executive*, 18(2), 35–39.

8. DTI (2003). *Innovation Report – Competing in the Global Economy: The Innovation Challenge*. Department of Trade and Industry – UK Government White Paper, London; and DTI (1998). *Our Competitive Future: Building the Knowledge Driven Economy*. Department of Trade and Industry – UK Government White Paper, London.

9. DTI (2004). *Critical Success Factors: Creating Value from Your Intangibles*. Department of Trade and Industry, London.

10. Blair, M. M. and Wallman, S. M. H. (2001). *Unseen Wealth*. Brookings Institution Press, Boston.

11. For more information, please see: Carmeli, A. and Tishler, A. (2004). The Relationships between Intangible Organizational Elements and Organizational Performance. *Strategic Management Journal*, 25, 1257–1278.

12. Marr, B. (ed.) (2005). *Perspectives on Intellectual Capital: Multidisciplinary Insights into Management, Measurement, and Reporting*. Elsevier, Boston.

13. Ibid. The terminology of assets is mostly linked with the accounting discipline and knowledge assets, and intellectual capital is often linked to knowledge management work.

14. See for example: Roos, J., Roos, G., Dragonetti, N. C. and Edvinsson, L. (1997). *Intellectual Capital: Navigating the New Business Landscape*. Macmillan, London.

15. For additional reading, see: Boisot, M. H. (1998). *Knowledge Assets: Securing Competitive Advantage in the Information Economy*. Oxford University Press, Oxford; Itami, H. (1987). *Mobilizing Invisible Assets*. Harvard University Press, Cambridge, MA.

16. See for example: Huselid, M. A. (1995). The Impact of Human Resource Management Practices on Turnover, Productivity, and Corporate Financial Performance. *Academy of Management Journal,* 38(3), 635–672; Carmeli, A. and Tishler, A. (2004). The Relationships between Intangible Organizational Elements and Organizational Performance. *Strategic Management Journal*, 25, 1257–1278; Huselid, M. A., Jackson, S. E. and Schuler, R. S. (1997). Technical and Strategic Human Resource Management Effectiveness as Determinants of Firm Performance. *Academy of Management Journal,* 40(1), 171–188.

17. See for example: Ghoshal, S. and Nahapiet, J. (1998). Social Capital, Intellectual Capital, and the Organizational Advantage. *Academy of Management Review*, 23(2), 242; Konno, N. and Nonaka, I. (1998). The Concept of "Ba": Building a Foundation for Knowledge Creation. *California Management Review*, 40(3), 40–54; Brooking, A. (1996). *Intellectual Capital: Core Assets for the Third Millennium Enterprise*. Thompson Business Press, London.

18. See for example: Parker, R. and Bradley, L. (2000). Organizational Culture in the Public Sector: Evidence from Six Organizations. *International Journal of Public Sector Management*, 13(2), 125–141.

19. Itami (1987), p. 23 (see note 15 above).

20. Marr, B., Neely, A. and Schiuma, G. (2004). The Dynamics of Value Creation: Mapping Your Intellectual Performance Drivers. *Journal of Intellectual Capital*, 5(2), 312–325.

21. See for example: Denison D. R. (1990). *Corporate Culture and Organizational Effectiveness*. Wiley, New York; Carmeli, A. and Tishler, A. (2004). The Relationships between Intangible Organizational Elements and Organizational Performance. *Strategic Management Journal*, 25, 1257–1278.

22. Nelson, R. R. and Winter, S. G. (1982). *An Evolutionary Theory of Economic Change*. Harvard University Press, Cambridge, MA.

23. See also: Clotier, L. M. and Gold, E. R. (2005). A Legal Perspective on Intellectual Capital. In: *Perspectives on Intellectual Capital* (Marr, B. ed.), pp. 125–136. Elsevier, Boston; Hall, R. (1989). The Management of Intellectual Assets: A New Corporate Perspective. *Journal of General Management*, 15(1), 53.

24. Hamel, G. and Prahalad, C. K. (1990). The Core Competence of the Corporation. *Harvard Business Review*, 68(3), 79.

25. See for example: Adams, C. and Johnson, D. (1998). *The Concise Blackwell Encyclopedia of Management*. Blackwell Business, pp. 624–625.

26. Collis and Montgomery (1997), p. 22 (see note 1 above).

27. Stalk, G., Evans, P. and Shulman, L. E. (1992). Competing on Capabilities: The New Rules of Corporate Strategy, *Harvard Business Review*, 70, pp. 57–69; Hamel, G. and Prahalad, C. K. (1990). The Core Competence of the Corporation. *Harvard Business Review*, 68(3), 79.

28. See for example: Rucci, A. J., Kirn, S. P. and Quinn, R. T. (1998). The Employee-Customer-Profit Chain at Sears. *Harvard Business Review*, 76(1), 83–97; Ittner, C. D. and Larcker, D. F. (1998). Are Nonfinancial Measures Leading Indicators of Financial Performance? An Analysis of Customer Satisfaction. *Journal of Accounting Research*, 36, 1–35.

29. For a debate on causal ambiguity and interrelatedness of resources, see: Lippman, S. A. and Rumelt, R. P. (1982). Uncertain Imitability: An Analysis of Interfirm Differences in Efficiency under Competition. *Bell Journal of Economics*, 13(2), 418; Dierickx, I. and Cool, K. (1989). Asset Stock Accumulation and Sustainability of Competitive Advantage. *Management Science*, 35(12), 1504; King, A. W. and Zeithaml, C. P. (2001). Competencies and Firm Performance: Examining the Causal Ambiguity Paradox. *Strategic Management Journal*, 22(1), 75.

30. Lev, B. (2001). *Intangibles: Management, Measurement, and Reporting*. The Brookings Institution, Washington, DC, p. 7.

31. Roos, G. and Roos, J. (1997). Measuring Your Company's Intellectual Performance. *Long Range Planning*, 30(3), 413.

32. Nobody really knows who invented SWOT analysis, although it was certainly being used by Harvard Business School academics during the 1960s. For more information, see also: Bryson, J. M. and Roaring, W. D. (1987). Applying Private Sector Strategic Planning in the Public Sector. *Journal of the American Planning Association,* 53, 9–22.

33. Since there are many different classifications and definitions of intangibles, it is important to stress that there is generally no right or wrong classification. Accountants, marketers, HR professionals and strategists have all long struggled with trying to categorize resources, especially intangibles, and their debate has yielded little insights. For an overview of this debate and a more detailed discussion of intangibles, see: Lev, B., Canibano, L. and Marr, B. (2005). An Accounting Perspective on Intellectual Capital. In: *Perspectives on Intellectual Capital: Multidisciplinary Insights into Management, Measurement, and Reporting* (B. Marr, ed.), pp. 42–55. Elsevier, Boston.

34. Due to strict confidentiality agreements, the name of this business cannot be revealed.

35. Göran Roos and Johan Roos have been instrumental in developing an understanding and mapping approach of resource interactions. For their insights on resource interactions, see for example: Roos, G. and Roos, J. (1997). Measuring Your Company's Intellectual Performance. *Long Range Planning*, 30(3), 413; Gupta, O. and Roos, G. (2001). Mergers and Acquisitions through an Intellectual Capital Perspective. *Journal of Intellectual Capital*, 2(3), 297–309; Pike, S., Roos, G. and Marr, B. (2005). Strategic Management of Intangible Assets and Value Drivers in R&D Organizations. *R&D Management*, 35(2), 111–124.

36. This approach was pioneered by Göran Ross. However, while Göran Roos et al., for example, use the navigator model to emphasize the influence and transformation from one resource to the next, here we assess the interdependence for creating a core competence. See for example: Marr, B. and Roos, G. (2005). A Strategy Perspective on Intellectual Capital. In: *Perspectives on Intellectual Capital – Interdisciplinary Insights into Management, Measurement and Reporting* (B. Marr, ed.), Chapter 2. Elsevier, Boston; Carlucci, D., Marr, B. and Schiuma, G. (2004). The Knowledge Value Chain – How Knowledge Management Impacts Business Performance. *International Journal of Technology Management*, 27(6/7), 575–590; Neely, A., Marr, B., Roos, G., Pike, S. and Gupta, O. (2003). Towards Third Generation Performance Measurement. *Controlling*, 15(3/4), 129–135.

37. See for example: Fink, A., Marr, B., Siebe, A. and Kuhle, J.-P. (2005). The Future Scorecard: Combining Internal and External Scenarios to Create Strategic Foresight. *Management Decision*, 43(2), 360–381.

38. See for example: Helfat, C. E. (2000). Guest Editor's Introduction to the Special Issue: The Evolution of Firm Capabilities. *Strategic Management Journal*, 21(10/11), 955; Pandza, K., Polajnar, A., Buchmeister, B. and Thorpe, R. (2003). Evolutionary Perspectives on the Capability Accumulation Process. *International Journal of Operations & Production Management*, 23(7/8), 822; Post, H. A. (1997). Building a Strategy on Competences. *Long Range Planning*, 30(5), 733–740.

Chapter 4

Mapping and Defining your Strategy

The previous chapters have given you an understanding of your stakeholder value proposition as well as of your core competencies and resources necessary to deliver it. Bringing these together allows you now to translate these into an integrated and coherent picture of your strategy. In this chapter, I will outline how a strategy can be visualized in a value creation map (VCM) and described in a value creation narrative (VCN). Together, they allow organizations to make their strategic plans explicit and communicate them to everyone within the organization in order for the strategy to be executed, challenged and refined. Questions addressed in this chapter include:

- Why do we need to map and define our strategy?
- How do we map our strategy into a VCM?
- How can we connect our inputs, outputs and outcomes?
- How do we describe our strategy in a VCN?
- How do we cascade VCMs in our organization?
- How have organizations applied these tools in practice?

Without an explicit understanding of strategy, strategic performance management will never be possible. Today, one of the biggest barriers to successful performance management is that strategy is often communicated in cryptic or incomplete ways, with the hope that employees will understand how it all fits together. In most cases they don't! Organizational strategy has its roots in military strategy, and strategy execution in organizations is often compared with an army fighting a battle. Two and a half thousand years ago, Sun Tzu, a wise Chinese general, wrote that an army that knows both its enemy and itself can fight a hundred battles without disaster.[1] He continues that once an army understands its enemy and knows its own strengths and weaknesses, it can lay out its plan. This is exactly what we are trying to do here.

Chapters 1–3 have given readers the insights of understanding the external competitive environment and the internal strengths and weaknesses of an organization. Now we need to bring them together in order to lay out a strategic plan and communicate it to everyone. Back to the army analogy which I have borrowed from Bob Kaplan and Dave Norton, we would expect an army general taking his troops into the battlefield to provide detailed maps of the terrain, telling field officers and troops exactly what the strategic objective is and how he envisages that it being achieved.[2] Without a detailed plan, it would be difficult, if not impossible, to clearly communicate a campaign. The likely consequence would be that the field officers and troops wouldn't really understand the strategic plan. However, in most organizations, leaders commonly communicate their strategy in cryptic and partial forms. It is not rare to see strategic business plans in public service and not-for-profit organizations that run into 35 or even 50 pages. These often include extensive lists of seemingly unrelated goals and targets without revealing how they all fit together; not surprisingly, in many instances they don't. It is also no surprise that employees and stakeholders are struggling to make sense of the strategy.

WHY MAPS AND NARRATIVES?

There are two primary functions of visual strategic maps and narrative strategy descriptions: (1) to ensure the strategy is integrated and cohesive and (2) to enable easy communication of the strategy. Our human brain interprets incoming information to create meaning. The Nobel Prize winners Roger Sperry and Robert Ornstein discovered that the brain is divided into two halves, or hemispheres, and that different kinds of mental functioning take place in each.[3] The left hemisphere operates sequentially and deals largely with 'academic' activities such as reading, arithmetic and logic. By contrast, the right hemisphere operates holistically and deals more with synthesizing and 'artistic' activities such as art, music, color and creativity (see Fig. 4.1). It is therefore easier for our brain to make meaning of complex information when it is presented in visual formats. Visual maps are processed in our right hemisphere, which is better equipped to deal with complex and holistic information. This is why a picture can be worth a thousand words.

Left-brain-focused people, who are strongly text oriented, could lose out on important insights to be gained from images depicted in pictures and diagrams.

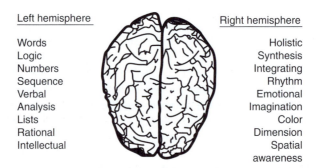

Left hemisphere		Right hemisphere
Words		Holistic
Logic		Synthesis
Numbers		Integrating
Sequence		Rhythm
Verbal		Emotional
Analysis		Imagination
Lists		Color
Rational		Dimension
Intellectual		Spatial
		awareness

FIGURE 4.1   Left brain, right brain.

The reverse may not hold true with the right-brain-focused people not necessarily being deprived of more linear and logical thinking. This may be due to our education, which already puts greater emphasis on 'left-brain thinking'. The early approach is then reinforced by the way we behave in business. Promotion of creative thinking into business activities can be seen as a means of counteracting our early and subsequent emphasis on linear thinking. Many great minds are believed to have used both parts of their brains. A good example is Leonardo da Vinci who, in his time, was one of the most skilled men in a wide range of disciplines such as art, sculpture, physiology, general science, architecture, invention, engineering, aviation and many more. As Mozart said, 'I can see the whole … at a single glance in my mind, as if it were a painting or a handsome human being.' It is the role of VCMs to focus everyone on 'the big picture.'

It is recommended that VCMs are accompanied by a VCN; here, a narrative means telling the story of value creation in writing. Dr Howard Gardner, Professor at Harvard University and author of *Leading Minds*,[4] believes that stories constitute one of the most powerful tools in business. Narratives have proven to be useful tools for organizations in communicating how they function, and especially about how intangible resources help to deliver value.[5] However, narratives not only provide additional contextual information, they also engage the logical left hemisphere of the brain and therefore ensure that both sides of the brain are involved in understanding strategy.

STRATEGIC MAPPING IN ORGANIZATIONS

Maps have long been used in strategic management to visualize complex relationships and knowledge.[6] Napoleon, for example, was a great advocate of their use. More recently, Robert Kaplan, Professor at Harvard Business School, and David Norton have made strategic mapping an essential part of their balanced scorecard (BSC) model, and thereby significantly contributed to the widespread usage of mapping tools in modern organizations.[7] Strategy maps are visual representations of the causal linkages assumed between strategic objectives in the following BSC perspectives:

- The *financial perspective* covers the financial objectives of an organization and allows managers to track financial success and shareholder value.
- The *customer perspective* covers the customer objectives such as customer satisfaction, market-share goals as well as product and service attributes.
- The *internal process perspective* covers internal operational goals and outlines the key processes necessary to deliver the customer objectives.
- The *learning and growth perspective* covers the intangible drivers of future success such as human capital, organizational capital and information capital including skills, training, organizational culture, leadership, systems and databases.

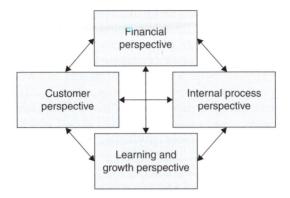

FIGURE 4.2   Traditional balanced scorecard template.

Initially, it was suggested to visualize a BSC in a four-box model (see Fig. 4.2). However, the classic four-box model is now outdated and has been replaced by a strategy map view. A strategy map places the four perspectives into a causal hierarchy to show that the objectives support each other and that delivering the right performance in the lower perspectives will help to achieve the objectives in the upper perspectives. For example, the objectives in the learning and growth perspective underpin the objectives in the internal process perspective, which in turn underpin the objectives in the customer perspectives. Delivering the customer objectives should then lead to the achievement of the financial objectives in the financial perspective. This causal logic is one of the most important elements of modern BSCs. It allows companies to create a truly integrated set of strategic objectives.

This visual representation of cause-and-effect relationships between distinct strategic objectives was first introduced into the BSC in 1996 and then extended in 2000. Today, Kaplan and Norton emphasize that strategy maps show how an organization will convert intangible resources into tangible outcomes.[8] The BSC strategy map template is depicted in Fig. 4.3.

A problem arises if public sector and not-for-profit organizations now take this generic strategy map template which was developed for commercial companies and just apply it mechanistically without changing its format to make it fit the public sector and not-for-profit environment. The overall goal of e.g. maximizing profits or delivering long-term shareholder value is not appropriate for public sector or not-for-profit organizations. Instead for most public sector and not-for-profit organizations finance is primarily an input resource that has to be managed as effectively and efficiently as possible, rather than an outcome that has to be maximized. With implementations of the BSC model at e.g. the City of Charlotte and the Boston Lyric Opera Kaplan and Norton have demonstrated that the tool can be successfully applied in public sector and not-for-profit organizations.

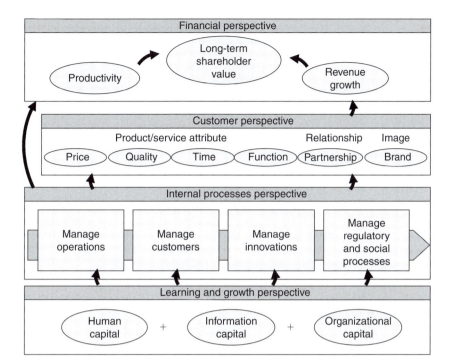

FIGURE 4.3   Kaplan and Norton's strategy map template.

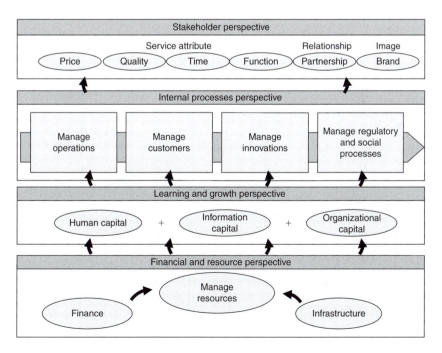

FIGURE 4.4   Public sector and not-for-profit strategy map template.

However, in order to make the BSC model relevant to public sector and not-for-profit organizations, a few changes have been proposed to the BSC template (see also Fig. 4.4):[11]

1. Move the financial perspective to the bottom of the template. The overall objective of most public sector, government and not-for-profit organizations is not to maximize profits and shareholder return. Instead, money and infrastructure are important resources that have to be managed as effectively and efficiently as possible to deliver the strategic objectives.
2. The main objective of public sector, government and not-for-profit organizations is to deliver services to their key stakeholders, which can be the public, central government bodies or certain communities. Therefore, the customer perspective is changed into a stakeholder perspective, which usually sits at the top of the template to highlight the key stakeholder deliverables and outcomes.
3. The two remaining perspectives will stay as they are. Any public sector, government and not-for-profit organization needs to build the necessary human, information and organizational capital to deliver its key processes in the middle of the map.

A tool was created in order to align the map with the organizational value proposition, the core competencies and the underlying resource architecture, as suggested in strategic management theory and as outlined in this book. The aim was to provide a simple strategy mapping template that does not require the shuffling around of the BSC perspectives. A so-called VCM builds on the important foundations laid by the BSC, and other performance management models, but ensures that the three key components of strategy are represented. A VCM provides an strategy mapping template that is truly grounded in the theory and practice of strategy. A VCM provides an approach that does not prescribe a certain business logic (which makes is appropriate for public service, not-for-profit as well as commercial organizations), provides a more comprehensive set of organizational resources and allows organizations to map their strategy into one integrated visual map.

WHAT IS A VALUE CREATION MAP?

A VCM is a visual representation of the organizational forward strategy. It brings together the three key elements of an organizational strategy, namely, its output stakeholder value proposition, its core activities, and its enabling strategic elements or drivers of performance. Let me define each of these in more detail and link them to the previous chapters:

- The *output stakeholder value proposition* (*or output deliverables*) answers the question of why an organization exists and what its roles and deliverables are. It identifies the key output stakeholders of the organization and describes what value (outcome or impact) the organization is delivering to them. It is mainly derived from the analysis of the core purpose and the output stakeholder requirements.

- The *core activities* are the vital few things an organization has to excel at in order to deliver the above value proposition. The core activities essentially define what an organization has to focus on and what differentiates it from others. Core activities are directly linked to the organizational core competencies and therefore derive from the assessment of the core competencies as part of the internal analysis.

- The *enabling strategic elements* (*or value drivers*) are the other strategic elements or objectives an organization has to have in place or has to deliver in order to perform its core activities and meet its output stakeholder value proposition. These enabling elements or value drivers are derived from the assessment of the resource architecture as well as the assessment of the input stakeholders and represent objectives linked to an organization's financial, physical and intangible resources.

These three components are then placed in relationships with each other and visualized on one piece of paper to create a completely integrated and coherent picture of the forward strategy. A VCM is therefore a visual representation of an organization's unique strategy at a specific point in time and it has a limited life span (usually 12 months, which is in line with the annual planning cycle of most organizations, but this can be shorter or longer depending on the dynamics in the external environment). As a consequence, there should never be two VCMs that are the same. The basic template of a VCM is shown in Fig. 4.5.

A VCM establishes a shared understanding and facilitates communication of strategy. Such shared understanding of the organizational strategy can then form the starting point for assessing, implementing and continuously refining the strategy.

The way the resources are visualized can vary depending on preferences, levels of understanding and available data. The most basic visualization is shown here, which does not show any causal relationships or individual interdependencies between the enabling elements. The fact that they are all in one box indicates that these different enablers or value drivers are interdependent and, as a bundle of enablers, support the core activities.

Organizations with a better understanding of their value drivers might want to move to more sophisticated mapping approaches within the VCM. The next level of sophistication is to highlight specific links between the individual value driver bubbles as outlined in Fig. 3.10. In Chapter 3, I discussed the fact

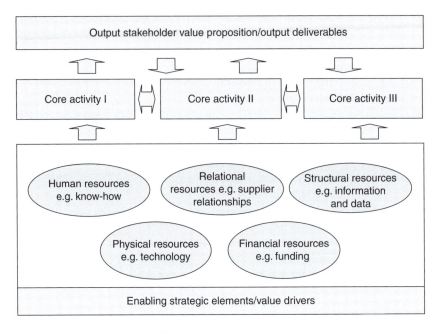

FIGURE 4.5 Value creation template.

that organizational resources depend on and dynamically interact with each other and the same is obviously the case for the resulting value driver objectives. To achieve a VCM with linked enabling elements, organizations can follow the methodology outlined in Chapter 3.

However, the most insightful visualization is to show cause-and-effect relationships in the form of dependence (also called 'influence') diagrams.[13] This will make VCMs truly operational and will lead to a deep and comprehensive understanding of the strategy. Figure 4.6 shows the different types of relationships between value drivers that could be shown in a VCM. The first is a simple arrow that indicates that there is a cause-and-effect relationship between the two value drivers – meaning that A leads to B. In order to add more detail to these maps, it is possible to show arrows with different widths to indicate stronger or weaker causal relationships. In the most sophisticated visualizations, organizations can differentiate the levels of dependence between the different resources. This allows organizations to see that, for example, the dependence of resource B on resource A is stronger than the dependence of A on B. For example, resource B might heavily depend on resource A, but resource A only moderately depends on resource B.[15]

Whereas a basic 'bundled' VCM, that is the one without the causal links, corresponds with the systems dynamics view,[16] which indicates that all

Relationship	Meaning	Example
Cause and effect:		
Enabler A ⟶ Enabler B	Enabler A leads to Enabler B	Good customer knowledge leads to good customer relationships
Influence/Dependence:		
Enabler A ⟷ Enabler B	Enabler A and B depend on each other	The right organizational culture depends on the right behavior, and vice versa
Enabler A ⇄ Enabler B	Enabler B strongly depends on Enabler A, and Enabler A depends on Enabler B to a lesser extent	Our processes depend heavily on our skills, and to a certain degree our skills depend on our processes

FIGURE 4.6   Possible causal relationships between resources.

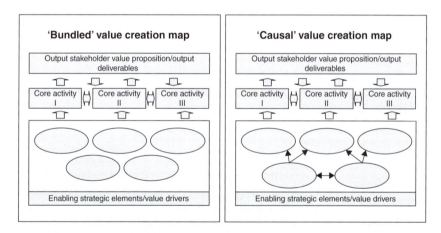

FIGURE 4.7   'Bundled' versus 'causal' value creation maps.

resources are interdependent and reliant on each other, a 'causal' VCM takes a more pragmatic view and visualizes the most important causal dependencies between the different resources. This makes VCMs easier to interpret and analyze, and also provides the possibility to verify and test the assumed causal relationships and interdependencies. Figure 4.7 shows both the basic 'bundled' view and the 'causal' view. Neither of them is right or wrong, and it sometimes depends on personal preferences or the worldview of manager as to which view an organization chooses. However, my experience with many public sector and not-for-profit organizations across the globe has taught me that causal VCMs are more powerful in communicating strategy. And even more importantly, they ensure that each value driver is in fact explicitly linked to the other strategic elements. This ensures that the strategy is coherent and integrated and that no elements have crept into the strategy for any other reason than being strategically relevant. The bundled view makes it a little too easy for an organization to just add elements to it which might only be there to please certain individuals or stakeholders. I feel that this is especially problematic in government and public service organizations which operate in a highly political context.

My recommendation is therefore to create causal maps that show the key interdependencies and causal relationships between the various elements, but at the same time acknowledging that all elements are to some degree interdependent and interlinked.

VALUE CREATION MAP VERSUS STRATEGY MAP

The VCM can basically be seen as a template to create a strategy map and in many instances the results will be identical. In Fig. 4.8, I have shown the traditional strategy map and the VCM template next to the tree I have used as an analogy throughout this book. On the left is shown the strategy map and on the right the VCM.

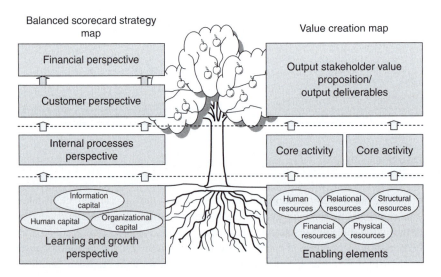

FIGURE 4.8   Value creation maps vs. balanced scorecard.

Because the BSC has been designed for commercial organizations whose key output stakeholders are shareholders, the financial perspective is at the top of the template. The other output stakeholders are customers who have to be satisfied in order to generate profits and shareholder value. This is why the customer perspective is placed below the financial perspective. The internal process perspective identifies where the organization has to excel in order to deliver the customer needs, and the learning and growth perspective identifies the intangible drivers of future performance and underpins the other perspectives.

This means that if someone follows the principles outlined in this book and created a VCM for a commercial organization, then the VCM would almost certainly mirror the BSC template. Shareholders and customers would be the key output stakeholders and would feature in the top box. The core activities would be outlined in the middle of the template and the key enablers and value drives, which would mainly be intangibles, are featured in the lower part of the map.

However, there are subtle but important differences:

- There is a danger in prescribing a specific business logic such as the causal relationship between the four perspectives in the BSC template. I see organizations which take the BSC model, accept it as it is without much reflection on their unique circumstances and then just populate each of the perspectives with objectives and indicators. The VCM on the other hand has no prescribed business model and therefore forces organizations to reflect on their situation to define their unique value creation logic.
- The VCM does not prescribe who the key output stakeholders are. Even though in most commercial organizations they are shareholders and customers, there can be others. Companies such as Johnson&Johnson put customers above shareholders, which would mean switching the top two perspectives in the Balanced Scorecard around. For most government and not-for-profit organizations, it would mean getting rid of the financial perspective at the top of the template and maybe changing the customer perspective into a community perspective. It is staggering how many public service and not-for-profit organizations blindly apply the strategy map template with the finance perspective at the top and then wonder why it doesn't work.
- In practice, the learning and growth perspective seems to be the least understood perspective of the BSC model. This is partly due to the name, which is not very clear or useful. Because companies associate learning and growth with their people, many companies rename this perspective into a 'people perspective' and only concentrate on elements such as training and staff satisfaction.[17] Drs Kaplan and Norton warn that by doing this organizations might miss other vital drivers of performance.[8]
- The strategy map template ignores any physical enablers such as infrastructure, buildings or land.

- The value driver perspective of the VCM is broader and includes performance drivers and enablers based on any organizational resource, be it financial, physical or intangible.

So overall, there is nothing inherently different between the VCM and the BSC or many other performance management frameworks; however, what I have done is stripped the various performance management models to their theoretical roots and then used this to build up the simple and generic template for the VCM, which can therefore be applied to any type of organization. Applying the VCM template does not require changing any perspectives around. Because it does not prescribe a value creation logic, it forces organizations to define their unique strategy but at the same time it provides enough guidance to make this a straightforward process.

HOW TO CONSTRUCT A VALUE CREATION MAP

A VCM can be drafted based on the information and insight from the analysis of the internal and external context of an organization. You can apply the tools outlined in Chapters 1–3 to extract the necessary information for a VCM (see Fig. 4.9).

Output Stakeholder Value Proposition

The key elements that will inform the output stakeholder value proposition are the purpose and visionary goals, together with the analysis of the output stakeholder requirements as well as the micro and macro environment.

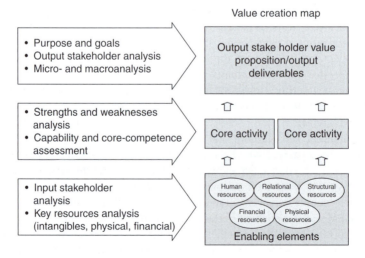

FIGURE 4.9   Creating a value creation map.

Reviewing this information should allow organizations to condense this into an output stakeholder value proposition. The stakeholder value proposition is usually in the format of a sentence or two. Here are some illustrative examples from both commercial companies and government and not-for-profit organizations:[18]

Belfast City Council:

The Council takes a leading role in improving quality of life now and for future generations for the people of Belfast by making the city a better place to live in, work in, invest in and visit.

Motor Neurone Disease Association:

The aim of the Motor Neurone Disease Association is to play a key role in ending the disease and until then, provide care and support for people affected by MND.

The leading pharmaceutical company, Novo Nordisk:

Novo Nordisk is leading the fight against diabetes. Defeating diabetes is out passion and our business.

The global courier firm DHL:

Delivering sustainable financial performance by making our customers successful through delivering high quality shipment solutions and superior customer service.[19]

A global insurance company:

Provision of sustainable financial security through excellent and customized trusted insurance covers for the global transport industry, together with value added service.

The medical Science department of a global pharmaceutical company:

To provide medical and scientific expertise into the drug development process in order to produce innovative, effective medicines that improve the health and quality of life of patients.

Core Activities

The core activities can be extracted from the analysis of the strengths (and weaknesses) as well as the assessment of the organizational capabilities and core competencies. Core activities are linked to the core competencies. Core competencies are usually formulated as statements of ability such as the 'ability to build 200 to 600 Watt motors', which is a core competence for Black and Decker. A core activity takes these and translates them into actionable activities on which the organization will focus in the next 12 months or so. The core activities refer to the few vital things you have to excel at and that define your organization and are instrumental to the delivery of the value proposition. I always suggest that a good number of core activities (the same applies to core competencies) is between two and four. If you identify more, then they are either not likely to be all 'core' or your strategy is not focused enough. Again,

I provide some examples of different core activities from commercial, public service and not-for-profit organizations:

- The Royal Air Force: Maintain and sustain combat ready equipment
- A global insurance company: Handle claims effectively and efficiently
- DHL: Understand the changing client requirements
- A government agency: Effectively engage with our stakeholders
- A global fast-moving consumer good manufacturer: Retail trade management
- Belfast City Council: Meeting the needs of local people through the effective delivery of quality and customer-centric services

Enabling Strategic Elements

The enabling strategic elements or value drivers are what is underpinning the entire strategy. They represent strategic objectives that are linked to the various resources (intangible, physical or financial), which need to be in place to deliver the strategy. One source is the input stakeholder analysis. If the input stakeholder analysis yields insights such as the contributions from employees, suppliers and partners, then these need to be reflected here. They usually map onto the intangible resources. However, instead of just listing the resources, they have to be translated into actionable objectives on which the organization will focus in the next 12 months or so. I suggest you use verbs and adverbs to define the value driver objectives to make them practical and relevant. As a rough guide, I suggest that you have between 5 and 15 enabling strategic elements or value drivers. If you have any less (which I have never seen) then I would question the depth of your strategic analysis and if you have more (which is common) then you might need to work a little harder to consolidate them and distil them down. Too many strategic elements make the maps cluttered and your strategy less focused and more difficult to explain, communicate and implement. Here are some examples of different value drivers from commercial, public service and not-for-profit organizations:

- Continuously improve and innovate our service delivery
- Optimize our supply chain
- Develop new products and services
- Forecast demand
- Grow and retain high-value customers
- Develop better relationship with key customers
- Improve our corporate reputation
- Increase customer confidence
- Build a strong brand
- Communicate effectively (internally/externally)
- Attract best-in-class people
- Acquire new resources/businesses

- Recruit new volunteers
- Leverage our IT infrastructure
- Provide relevant material on our Web site
- Train and develop our staff in the area of X
- Recognize the contribution of our people
- Engage with our suppliers
- Manage financial budget efficiently and effectively
- Reduce costs of service delivery
- Create a performance-driven culture
- Share and live our values

If you feel that a lot of strategic thinking and analysis has already been done in your organization and you don't feel the need to complete all the different analyses outlined in Chapters 1–3, then you can shortcut this process by interviewing key people in your organization and reviewing existing analysis reports and strategy documents. Interviewing the key people allows you to extract everybody's view of the strategy, which can then be supplemented with relevant documents and analysis to bring this together into a draft VCM. The fact that everyone is interviewed individually is important as it overcomes some of the political barriers and ensures that everybody has equally and fairly contributed to the process. The interviews and subsequent analysis are best conducted by an experienced external facilitator. This facilitator can then analyze the different data sets from the interviews and documents and draft a first version of a VCM.

Once the first draft is created, it makes sense to bring everyone together in a workshop where this map is then reviewed. If this workshop is facilitated well, then the interactions as well as the immediate questions and feedback should lead to consensus about the forward strategy. There are advantages if this workshop is again facilitated by an experienced facilitator and, if possible, someone external to the organization. This is especially recommended if there are any dominant participants who might be able to impose their view of reality on other participants. In the workshop, some linkages might be deleted and others might be emphasized, some elements might be consolidated or disaggregated and others might be added or deleted.

A good practice is then to draw the finalized and agreed map and mail it out to all participants after the workshop for consideration and comments. This gives people some time to think about the final map and compare it with the reality of everyday business. Feedback can then be collected and, if necessary, a final workshop can be arranged to agree on the final layout of the map.

I have facilitated this process in hundreds of organizations and most people find this process extremely stimulating and engaging, and we have always reached an agreement on the final map during the workshops. The more people you engage in this process, the better it is. However, there are a certain minimum number of interviews, and there is a point where more interviews will not yield anymore significantly new information. From experience, the ideal number of people to interview is between 10 and 20. I would usually suggest involving the entire top management team, some selected middle managers and, if deemed necessary, a few frontline employees and representatives from external stake-holders. Any additional interviews are usually conducted for 'political' purposes and to ensure the inclusion of all those interested. However, around 15 interviews will give you a robust enough data set to draft the first version of the VCM.

When I have facilitated this for organizations in the past, they have some-times insisted on larger scale data collections. In DHL's case, for example, we conducted surveys of all employees and of their 300 key customers to explore everybody's views on the key output deliverables, competencies and key resources. In addition to that we conducted a set of interviews with the sen-ior management team as well as a selection of middle managers and frontline employees. Once the data was analyzed, we found that the interviews on their own would have provided sufficient information to draft the VCM and little additional information was collected from the surveys.

I do not recommend that you try to build these maps in a workshop envi-ronment. The reason for this is that you always get one or two people who are more powerful or more articulate than others and, as a consequence, they can easily bring a bias to the map. People might nod during the workshop but feel that their views haven't really been taken into account. Workshops should only be used to discuss the draft map after the interviews.

When I and my colleagues from the Advanced Performance Institute help clients to design their VCMs, we use a semistructured interview process in which we ask questions to explore the three key areas of strategy. Below I have given some examples of questions that could be used in interviews in order to understand the three main components of the VCM:

- Why does this organization exist? What is its main purpose? What do you do, and for whom? What is your value proposition?
- What do you have to be really good at? What do you have to excel at in order to deliver your value proposition? What are your core competencies and core activities as an organization?
- What are the underpinning enablers and drivers of performance? What key resources do you possess or need as an organization? What about people and their skills and knowledge? What about your infrastructure (IT etc.)? What key relationships do you need? What about your brand, your image, patents, organizational culture, your processes and practices etc.?

It is also always good to give the VCM a 'corporate' feel by using familiar language, colors and formats. Use formats, symbols and colors that are used in your organization so that they are comfortably accepted by everyone.

Organizations continuously evolve and change. The external and internal environments evolve and so the forward strategy needs to evolve too. This means that the value creation map needs to be revised on a regular basis in order to ensure that it reflects a current view of the strategy. How often these revisions take place depends on the speed of change in the sector or industry the organization is part of. It is usually a good idea to align the revisions of the VCM with the strategic planning cycles and, for many organizations, an annual revision is adequate. However, in some emerging or fast-moving industries this revision cycle can be accelerated.

DEFINING THE ELEMENTS ON YOUR MAP

All of the above examples of strategic elements are taken from different VCMs of clients I have worked with. Some have been generalized to facilitate understanding for readers. However, I would encourage you to ensure that your strategic elements are as specific as possible. For example, instead of using generic descriptions such as 'Train our people', it is usually more effective if you can be more explicit and say, for example, 'Train our ground staff to improve aircraft turnaround'.

I realize that this is sometimes difficult or impossible to do if you create a corporate VCM for a large organization. This is why each of the elements on the VCM needs to be defined in a bit more detail.

I recommend that each element's core activities and enabling strategic elements are defined beyond just the short name. This can be achieved in a one- to three-sentence description of what these elements mean.

WHAT IS A VALUE CREATION NARRATIVE?

A VCN is defined as a concise piece of written work that describes the organizational strategy and tells the story of how that organization intends to create value by specifying its value proposition, required core competencies and key resources. A VCN is there to accompany a VCM and provide additional contextual information and allow organizations to explain the chain of events in a story format.[20] Pictures and stories are the way we, as human beings, have communicated over thousands of years and our brain is predisposed to absorb information in narrative form.

The format of a VCN is not prescriptive. It very much depends on organizational preferences and should be aligned with the corporate look and feel. However, a VCN should be between 500 and 1000 words long. The story should be clear and readable, and written in a conversational style.[21] It is recommended that jargon or technical terms which readers of the narrative might not be familiar with are avoided as this may interfere with their understanding. At the same time, it is important to use language and terms that are usually used and understood within the organization. Clarifying the strategy in a concise narrative is a powerful way of clarifying the organizational strategy. I have provided a number of VCN examples in the case studies that follow (see e.g. Belfast City Council and The Home Office).

HOW TO CONSTRUCT A VALUE CREATION NARRATIVE

A VCN provides contextual information of the strategy that is visualized in a VCM. The VCM is therefore the starting point for writing a VCN. Based on the map, an individual or a small group of people can produce a draft version of the narrative. This draft is then circulated to a wider group of people, usually the senior management team, who then review the document and submit comments and suggested edits. These are then collected by the individual or group that produced the draft to create the final version of the narrative. Alternatively, and possibly for best results, the draft version of the VCN is circulated and subsequently a group session is arranged with the senior management team to edit the report in an interactive workshop. However, this process can produce overly long narratives – please remember that it *must* be concise.

CASCADING VALUE CREATION MAPS

A frequently raised question is when and how to cascade a VCM. A VCM is supposed to be a management tool that clarifies the strategy and therefore creates a shared identity and engages people in assessing and evolving this strategy. If this strategy is too distant and hence too abstract to people, it fails in its role. The employees need to relate to the elements on the map – it must reflect their reality in order for them to identify with it. It therefore becomes clear that a corporate VCM of, for example, a large and diverse organization will not be meaningful for someone working as a middle manager in one of the many departments or business units. It becomes too generic, too abstract, and similar in many ways to corporate mission statements that are often no more than a set of well-meaning words that have little operative value for people working further removed from corporate centers. For that reason, a VCM has to be cascaded and translated to a meaningful local perspective.

How many VCMs an organization needs depends on its size and diversity. Let's take one of the commercial companies I have worked with, for example, Royal Dutch Shell plc. Shell is a group of diverse companies that includes 'Exploration and Production', responsible for finding and producing oil and gas; 'Renewables', building commercial-scale wind parks and selling solar photovoltaic panels; 'Shell Trading', trading of crude oil, refined products, gas, electrical power; and 'Shell Global Solutions', providing business and operational consultancy, technical services and R&D expertise to the energy industry worldwide, among other corporate entities.[22] The same would be the case for large government and not-for-profit organizations. Just think of the various departments of the US Army or federal government departments such as the Department of State or the Department of Education. It is clear that all of these have large subdepartments and business units that have different stakeholder value propositions, unique core competencies and very dissimilar resource architectures. As a consequence, each of the different subdepartments or business units would require their own VCM. However, even though the maps might have different components, there should be elements of connectivity between them (sometimes referred to as the 'golden thread').

For more homogeneous organizations, a VCM template can be used. Take, for example, commercial companies such as Burger King or Hilton Hotels. Their individual restaurants or hotels might have much the same generic value propositions as well as an almost identical set of core competencies and resources but, potentially, substantially different product or service ingredients and pricing arrangements that reflect different customer expectations in different parts of the world. This means a generic template can be developed which then allows individual departments and business units to translate this into their VCM.

When I did strategic performance management work for federal and state governments, police forces or air forces, we designed a corporate map in the way described above and then created a template to facilitate the cascade. In the case of the air force and the police force, we created a generic template of overall output stakeholder value proposition, core activities and key resources. This worked well for the police and the air force where they have similar air bases or police stations. Each air base or police station helps to contribute to the overall output value proposition of their mother organizations (the police force or the air force). However, the different air force bases have different functions, one might host surveillance aircrafts and others might host fighter jets. This means there needs to be a scope to amend the output value proposition to make it relevant to the various air bases or police stations. The core activities should be similar too. Again, we created a template that included the general core activities for police stations or air force bases, but then gave the stations and air bases the freedom to verify them and to potentially translate them into slightly different and in some cases additional core activities. Also, the resources required to perform well should be fairly similar across the different police stations and air bases and so we created a template that included the key resources. The different police stations and air bases took this template and again discussed it and translated it into relevant objectives (enabling strategic elements) for their station or base. I therefore recommend that if you are using templates to cascade the VCM then it is important not to impose the final activities and objectives that will go on the cascaded maps but to provide a generic framework that guides and aids the development of a relevant local map but doesn't prescribe specific tasks and objectives. The resulting local VCMs might look similar at first glance, but some of the core activities and value drivers should be different and additional core activities and value drivers might appear of the cascaded maps.

Also, each organization usually has a diverse set of functional business units such as operations, HR, marketing, finance, logistics, IT etc. These functional business units contribute to delivering the overall value proposition of the business as a whole, but each of them has a value proposition of its own.

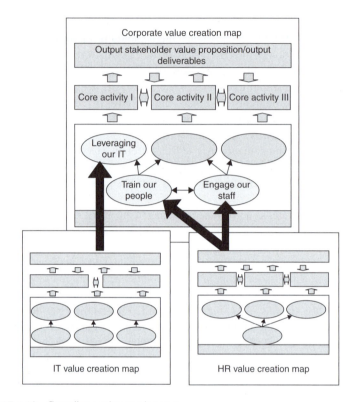

FIGURE 4.10   Cascading a value creation map.

Various books and articles have been written about how to strategically align functions, such as HR or IT, with the overall strategy of the business.[23] It is important that all functional business units understand how they contribute to the overall value proposition. In the case of an IT function, its key contribution might be to one of the value drivers on the corporate VCM, for example, an innovative technology base, whereas the HR function might be contributing to various 'bubbles' (see Fig. 4.10).

If a VCM is to become a tool that facilitates strategic decision making and learning at every level of the organization, then it is critical that the information provided is relevant to its people. The field of anthropology has found that humans have a limit to the number of people with whom we can retain a social relationship. This limit is likely to be linked back to the clan sizes of our hunter and gatherer ancestors. Research evidence shows that it is hard, if not impossible, for us humans to share an identity with more than about 150 people.[24] Experience has shown that this number is a good benchmark for the cascade of VCMs, in terms of content rather than structure.

The best way to cascade a VCM is to use an internal or external facilitator; preferably someone who has been involved in the creation of the organization's overall (or master) VCM. The process is the same as the one describing the construction of a VCM above. However, the external analysis will be reduced to a minimum as it only consists of identifying the key stakeholders and their needs. This can often be derived from the existing organizational VCM. If this is not the case, then some of the external analysis tools can be used to determine how the business or business unit helps to deliver the overall value proposition. Once the value proposition is clarified, the internal analysis will provide the data to create a cascade of the VCM.

VALUE CREATION MAPS AND NARRATIVES IN PRACTICE

The approaches outlined above have been implemented by hundreds of organizations around the globe including global enterprises such as Astra Zeneca, BP and DHL as well as many governments, public service providers and not-for-profit organizations including federal, state and local government institutions, universities, police services, fire services, schools, hospitals, charities and many more. They have proven successful for a very diverse group of organizations from large international blue-chip corporations to very small- and medium-sized companies. It is important to highlight, however, that every organization takes its own journey, with its own interpretations of the tools and techniques. Below I have outlined some illustrative case studies that demonstrate how leading organizations have translated these concepts into reality and applied the tools in practice. They serve the purpose of being real-life examples that I hope might provide some guidance, but they will never provide templates you can simply copy to create your own VCMs and narratives. Every context and every strategy is different. VCMs are unique descriptions of an organization's forward strategy at a given point in time. For more case examples, please visit the resources section of the Advanced Performance Institute's Web site: www.ap-institute.com.

Case study: Belfast City Council[25]

Belfast City Council is the largest local authority in Northern Ireland providing local political leadership and a range of services including refuse collection and disposal, street cleansing, building control and environmental health, community development, indoor and outdoor leisure, parks and recreational facilities, and support for the arts, tourism and economic development. The Council area has a population of around 269 000 and employs more than 2600 people. Decisions on how the Council is run are made by 51 elected councillors whose role is to make sure the views of the people of Belfast are reflected in the way services are provided.

The chief executive, along with councillors, chief officers and heads of service agreed that the Council needed to develop and improve if it were to become a modern, twenty-first-century local authority, and that they needed to be in a position to respond to the challenges that lie ahead. For this reason, Belfast City Council decided to develop and implement a state-of-the-art strategic performance management and measurement approach. Here, I outline how the Council used the VCM to clarify and agree on their corporate strategy.

Belfast City Council wanted to be in a position to improve quality of life in the city by improving both service delivery and the Council's civic leadership role. To achieve the necessary improvements to make this happen, it was agreed that the Council should focus on the following areas of work:

- *Governance*: Enabling more efficient and better decision making
- *Performance management*: Providing support and resources to help get results
- *Resource allocation and planning*: Matching resources to priorities
- *Customer focus*: Putting people first
- *People management*: Building capacity across the organization

The agreed process for taking this improvement forward was a 'strategic performance management' method,[26] which involved developing a VCM for the organization as a whole as well as individual VCMs for services, to replace traditional service-level business plans.

The diverse political environment in Northern Ireland means that there are six political parties in Belfast City Council with no single party in overall control. This was also reflected at officer level, resulting in a fragmented culture with low corporate levels of solidarity creating pronounced silo mindsets. Therefore, any process to define a strategy for the organization needed to be inclusive in order to agree on one strategy for the city. It was recognized that the absence of an agreed and clearly defined strategy would severely jeopardize future management and decision-making processes in the council. The plan was to use the VCM in order to bring together the different views and to clarify and visualize the strategy of the organization.

The VCM for Belfast City Council was designed based on the input of elected members and senior officers as well as a review of existing strategy- and planning-related documents. The key steps taken in designing the VCM for Belfast City Council are outlined below:

1. Scoping: First, the project was scoped and planned. As part of this, it was decided whom to involve in the strategy development process. In order to get a broad and balanced view across the council, it was decided to involve all chief officers, heads of services and elected members from all parties.
2. Collection of data: Individual in-depth and semistructured interviews were conducted with all chief officers, heads of services and elected members. In addition, observation data and document reviews (e.g. business plans, strategy reports etc.) were collected and used to supplement the interview data.
3. Creation of a value creation map: The interview data was analyzed to extract themes, constructs and insights to design a draft VCM. A feedback workshop was used to present the draft VCM to senior officers and elected members. Feedback was collected during the workshop, which led to minor amendments to the map. Further feedback was collected in the weeks following the workshop, which led to the final version of the VCM. In a subsequent meeting, the

new strategy captured in the VCM was agreed upon by both officers and members. For the first time the council had an agreed and clearly defined strategy outlining its value proposition, core competencies and enablers of future performance.

4. Creation of element definitions and narrative: Once the VCM had been created, additional information was required for each of the elements on the map as to what they really mean. For that purpose, a one or two paragraph definition was created for each element to provide further detail. This was achieved in a series of meetings and workshops. A smaller project team was used to take this part forward and drafted the definitions in close collaboration with the relevant senior officers. Feedback loops were used to ensure chief officers and members were informed about the progress and were able to provide feedback and suggestions.

The Council Improvement Board, the Chief Officer Management Team and the Heads of Service Forum, together with the Core Improvement Team, refined and finalized the corporate VCM, which was subsequently agreed and signed off by the Policy and Resources Committee. Figure 4.11 depicts the corporate VCM for Belfast City Council and the following narrative describes it.

Belfast City Council's VCN:

The main purpose of Belfast City Council is to help improve quality of life for the people of Belfast now and in the future by making the city a better place to live in work in, invest in and visit. To do this we must be good at two things. The first is to provide strategic leadership and direction and work with others to shape, develop and manage a shared city. We will also continue to meet the needs of local people by providing a wide range of quality and accessible services. As a Council we have identified a number of key areas we will focus on to achieve our goals.

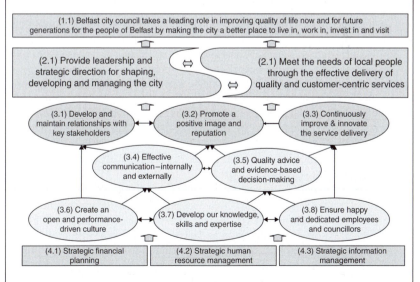

FIGURE 4.11   Value creation map for Belfast City Council.

We have a wide range of key stakeholders including European, central and local government, the voluntary and private sector, public agencies, citizens, funding bodies, neighborhoods, media, politicians, academia and professional bodies. We recognize the need to work well with all of these stakeholders if we are to improve the co-ordination of service planning and delivery and assist with the implementation of the Review of Public Administration (RPA). To do this we will build public confidence by promoting a more positive image of the Council among the media and by supporting Councillors in their work to represent the organization and the City. This will require improving officer/Councilor relationships to build trust and facilitating more two way dialogue among employees and stakeholders. We will be clear about what our priorities are and will effectively communicate and listen in an open and transparent way.

To achieve these improvements we will create an open, performance driven culture built on trust, where performance is discussed openly and used to help the organization learn and improve. Everyone will know what we want to achieve and how they contribute to this, in an environment where performance counts, is valued and is at the heart of everyone's job.

We will also identify the skills and expertise necessary to be a successful organization and Councillors and employees will work together to develop skills and improve how knowledge is shared across the organization. BCC will be a place where people are happy and motivated to do a good job. All decision makers will have access to the right information and expertise to allow them to make informed decisions. This will involve improving our structures to ensure all decisions are transparent, made at the right level and are acted on quickly.

All parts of the Council will work to bring about innovative improvements in service delivery for the benefit of our customers. To do this we will re-align resources, make better use of technology, bring about more joined-up working and encourage and reward innovation and improvement at all levels.

I would like to emphasize that this strategy is a representation of Belfast City Council's focus at the point of writing this book. However, a strategy is not a static document; it has to be continually reviewed as new information and updates are used to inform its future content. A good practice is to review and revise the strategy as part of the annual planning cycle to ensure the VCM keeps evolving with the organization. Belfast City Council has just gone through its first review cycle and has updated its corporate strategy (for the latest version, please see www.ap-institute.com).

Case study: The Royal Air Force[27]

The Royal Air Force of the United Kingdom (RAF) has 50 000 service and civilian personnel and more than 500 aircraft. The RAF supports operations in the Gulf region, Kosovo and Afghanistan as well as maintaining an RAF presence in Cyprus, Gibraltar, Ascension Island and the Falkland Islands. Its key peacetime responsibility is to maintain the required readiness levels of its forces (e.g. the Harrier, Globemaster, and Sentinel aircraft and their crews) in support of the requirement to operate as an expeditionary air force.

Performance management plays a key role in the RAF. The performance management approach within the RAF consists of a series of interdependent hierarchies of performance indicators to inform senior commanders on the current and forecast readiness of their forces to meet the range of war scenarios agreed with the government. The performance indicators are reported in four main perspectives of management, namely, resources supplied, processes undertaken, outputs delivered and enhancements for the future.

To sustain its activity, the RAF is organized into four layers of management: service, command, group and station. The aircraft and their crews are based at some 30 stations from which the aircrew, supported by ground staff, train and operate. Thus, the best level of knowledge on the readiness, current and forecast is at station level. For this reason, stations form the backbone of the performance management reporting process, supplying the raw data supplemented by the local commander's judgment on the situation. This information is vital to commanders higher in the command chain, informing them of the situation on the ground so that they can provide the most effective guidance and direction, while deploying the available resources most effectively.

I have worked closely with the RAF Air Command to cascade the overall performance management approach into relevant VCMs for the various RAF stations. The first task was to convince the different station commanders that a local VCM was useful. While the requirement to report to higher management was accepted, the existing task of collecting and reporting performance indicators was often done grudgingly by station staffs. The nature of what had to be reported was often at such a low level of granularity that it rarely provided useful information for the management of stations. The underlying problem was that stations couldn't make the connection between the corporate reporting they had to do and the local strategy they were following.

It was recognized that there was a significant risk that this situation could lead to local strategies being out of alignment with higher level goals. Greater connectivity needed to be achieved between local strategies and corporate strategies. To achieve goal alignment as well as local relevance, RAF stations embarked on a journey of cascading the overall RAF strategy into local performance management systems including local VCMs.

In order to create the local maps, we conducted individual interviews with each of the station executive in their own office. This was useful, since on a busy frontline station a number of executives would have their offices close to the taxiways and operational activities and there was often useful information to be gained from seeing the executive operate 'in situ'. Based on the information collected from the interviews as well as observations and the review of relevant documents, a picture of the station emerged. The essential resources on which the station relied (e.g. people, equipment, runways and buildings) were largely evident. There were also several obvious core activities, such as flight training, servicing of aircraft and administrative support, which needed little thought. However, the importance of maintaining fighting spirit and cohesion across the unit meant that there were a number of intangible, but nonetheless essential, value drivers that the station needed to be competent at. The emerging picture was translated into a VCM charting the enabling strategic elements flowing to the core activities to the delivered output to achieve the overall mission.

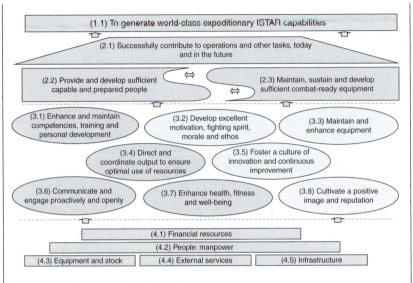

FIGURE 4.12   RAF station value creation map.

The goal was always to represent the essence of a station on a single A4 page. For the RAF context, the resultant diagram was termed the strategic map. The draft strategic maps were then subjected to rigorous review during a presentation given to the station commanders and their executives. While there were differing views on the key interdependencies and the relative importance of core activities, an agreement was reached on the essential components of the map. Once the map had been agreed in principle, an associated table was generated, containing an explanation of the intended scope of each element of the map. This was necessary to ensure a common understanding of the elements of the map and proved particularly important as a number of the map's elements crosscut over organizational boundaries and conventional processes.

Figure 4.12 outlines the VCM for one of the RAF stations. Overall, this station (RAF Waddington) exists to generate world-class expeditionary intelligence, surveillance, target acquisition and reconnaissance (ISTAR) capabilities. The station has three core activities or things the station has to excel at, namely, to successfully contribute to operations and other tasks, today and in the future; to provide and develop sufficient capable and well-prepared people; and to maintain, sustain and develop sufficient combat-ready equipment. The station agreed on eight drivers of performance that would enable the station to continue delivering its objectives. These performance drivers are to: enhance and maintain competencies, training and personal development; to develop excellent motivation, fighting spirit, morale and ethos; to maintain and enhance equipment; to direct and coordinate output to ensure optimal use of resources; to foster a culture of innovation and continuous improvement; to communicate and engage proactively and openly; to enhance health, fitness and well-being; and cultivate a positive image and reputation. Other key resources were identified as money, people, equipment and stock, external services and infrastructure.

Case Study: the Home Office[28]

The purpose of the Home Office, a key central government institution in the United Kingdom, is to work with individuals and communities to build a safe, just and tolerant society, enhancing opportunities for all. In such a society, rights and responsibilities go hand in hand, and the protection and security of the public are maintained and enhanced. This involves reducing crime and the fear of crime, including combating terrorism and other threats to national security; ensuring the effective delivery of justice; regulating entry to and settlement in the United Kingdom effectively in the interests of sustainable growth and social inclusion; facilitating travel by UK citizens; and supporting strong and active communities in which people of all races and backgrounds are valued and participate on equal terms. The latter can be achieved by developing social policy to build a fair, prosperous and cohesive society in which everyone has a stake.

This case study is based on the work of the Immigration and Nationality Directorate (IND), one of the Home Office departments that, together with the Department for Constitutional Affairs (DCA) and UK visas, will deliver the government's asylum and immigration strategy. The project was part of a wider initiative of the government to improve performance management.

IND's *Value Creation Narrative*

The high-level objective, as set out in the published Home Office strategic plan and the vision statement in the DCA 5-year strategy, is that migration is managed to the benefit of the United Kingdom, while preventing the abuse of the immigration laws and of the asylum system.

The key output deliverables that IND needs to deliver in order to achieve the high-level objective are effective control, support of legal migration, value for money and community cohesion. This involves continuing to encourage legal migration, which supports the UK economy, while remaining firm against abuse, and increasing value for money with demonstrable year-on-year efficiency gains across the organization. Value for money here is a combination of (i) doing the same for less, that is reducing costs; (ii) increasing the amount it achieves with the same money; and (iii) using money more effectively. It also involves building strong, cohesive communities for which it is important that long-term migrant workers and genuine refugees are swiftly integrated into society through settlement and citizenship. Effective integration will empower migrants to achieve their full potential as members of British society and thus help to build cohesive communities.

The key activities that help IND to deliver its output deliverables are continuous process improvement and effective stakeholder management. Stakeholder management focuses on international collaborations, effective delivery partnerships, responsiveness to customers and public confidence. Process improvement focuses on improved quality and productivity, simplified and joined-up processes and effective resource management.

To achieve the above, IND needs to develop as an organization and build the right resources for the future. Achieving its core activities and output deliverables is based on the right human resources, the right technology base, the right knowledge management processes, as well as continuous organizational development.

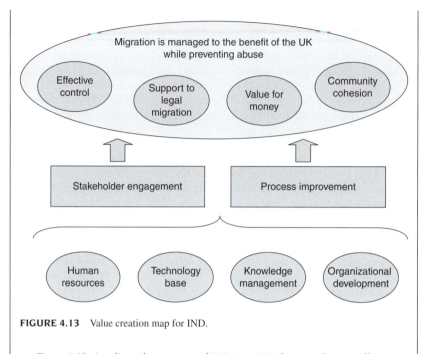

FIGURE 4.13   Value creation map for IND.

Figure 4.13 visualizes the strategy of IND in a VCM format. The overall output deliverables, together with the high-level objective, are at the top. Below are the two core activities IND needs to excel at in order to deliver its proposed value proposition. At the bottom of the map are the key drivers that IND needs to manage in order to be successful.[29]

Case Study: The Scottish Intellectual Assets Centre[30]

The Intellectual Assets (IA) Centre is the operational arm of Scottish Intellectual Asset Management Limited – a public sector organization developed as a joint subsidiary of the two development agencies of Scotland, Scottish Enterprise and Highlands and Islands Enterprise. The Centre is based in Glasgow and is sponsored by the Scottish Government and part-financed by the European Union. It brings together a range of expertise to help organizations realize their potential through managing their IA. The IA Centre employs a highly professional team of people, whose responsibilities reflect the company's commitment to the highest standards of efficiency, business effectiveness and their key strategic role in Scotland's economic future.

The Centre was established in 2003 to develop awareness of IA and to increase the level and effectiveness of IA management and exploitation amongst Scottish businesses, public sector and voluntary sector organizations. As part of its portfolio of activities, the IA Centre undertakes structured programs of interventions

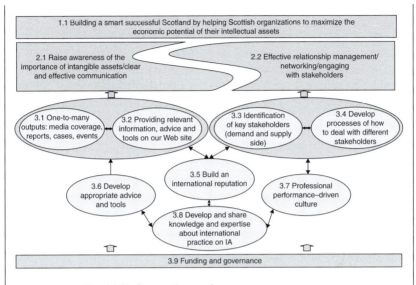

FIGURE 4.14   Scottish IA Centre value creation map.

aimed at improving the management of IA by small- and medium-sized enterprises (SMEs) in Scotland. To achieve this, the IA Centre acts as a catalytic and evangelizing body that is setting out to develop a high profile as a Centre of Excellence in IA issues. It seeks to be seen as a resource that complements and adds value to the activities of public and private sector intermediaries who support businesses and economic development. The Centre has a key role in assisting in the development of public policy and the economic and innovation development framework in Scotland and acts in an advisory and consultative role in this regard.

However, the IA Centre is itself an organization that heavily relies on intangible assets. It has to manage and measure its own non-financial value drivers. This case study outlines how the IA Centre, in collaboration with the Advanced Performance Institute, visualized their strategy with all its intangible value drivers to better manage their intangibles going forward.

The IA Centre wanted to clarify its strategy and identify the critical intangibles and the way they help to drive competencies and value creation. The VCM for the IA Centre was designed on the basis of the input from all staff, managers and directors of the Centre as well as a review of existing strategy- and planning-related documents. The final VCM that staff and managers of the IA Centre agreed on was signed off by their board in a board meeting. Figure 4.14 outlines the final VCM for the IA Centre. Below I provide the definitions for each of the elements (using the terms created by the IA Centre at the time).

Output stakeholder value proposition

● The overall objective of the IA Centre is to help build a smart successful Scotland by helping Scottish organizations to maximize the economic potential of their IA.

Core activities

- One of their core competencies is raising awareness of the importance of intangible assets. This involves stimulating activities in the market as well as clear and effective communication.
- Another core competence of the Centre is the effective relationship management and engagement with stakeholders. This goes hand in hand with the first core competence as it is only through leveraging their networks that they are able to raise the appropriate awareness and make effective interventions.

Key enablers or value drivers

- To provide one-to-many outputs – especially appropriate proactive quality media coverage, reports and events – with the aim of furthering people's understanding and to build awareness.
- To provide information, in particular on our Web site as well as face to face. This is about communicating and helping our customers to understand the importance of IA. It includes, in particular, proactive, managed one-to-many communication, but also includes one-to-ones.
- To identify our key stakeholders and get a hold on whom we should build relationships with. We need to prioritize stakeholders on both demand and supply side.
- To develop processes of how to engage with our key stakeholders from both supply and demand side. This is about having the right means to build the relationships and the appropriate terms of engagement.
- To build a national and international reputation. A good reputation helps us to better communicate and build better relationships. An international reputation helps us to gain a national reputation, especially on a political level.
- To develop the appropriate advice and tools. The advice, tools and methodologies developed by the Centre are part of an innovation system, where constant review is required in order to meet the emerging and changing needs of Corporate Scotland.
- The culture is one in which people are happy to be managed by objectives (team objectives, project objectives) and like to work together to deliver them. It is a supportive and evolving organization with which people identify and are passionate and share a vision.
- To continuously develop and share their knowledge. A great asset is the fusion of perspectives in the area of IA which need to be shared and brought into one cohesive body of knowledge. It also includes discussing IA on an international stage and bringing new insights back.
- To secure sustained funding for the organization and to ensure that it complies with the public sector regulatory framework and legislation.

Case Study: Insurance Mutual (TT Club)[31]

The TT Club is a mutual association and a leading provider of insurance and related risk management services for the international transport and logistics industry. The company has its global headquarters in the city of London, the

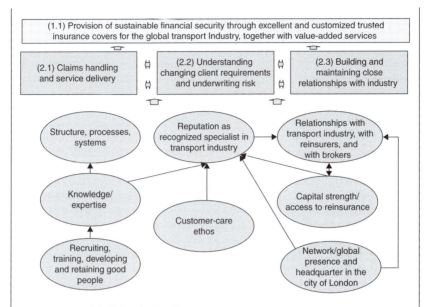

FIGURE 4.15   TT Club value creation map.

central hub for insurance firms, but has 20 office locations around the world. Its customers range from the world's largest shipping lines, busiest ports, global freight forwarders and cargo-handling terminals to smaller companies operating in niche markets. Since its inception, the TT Club has grown steadily in terms of premium income, at an average rate of 10% per annum for the last 20 years.

As a mutual association, the TT Club is owned by its policyholders (members) and does not work for profit. This means not only that the members decide how the Club is run, but also that all the funds of the TT Club are used for the benefit of the members. Its income is derived from premiums, service fees and earnings on invested funds. Outgoings are limited to claims payments, administrative expenses and reinsurance costs. Customer loyalty has been an essential factor in this growth. Indeed, 90% of its customers renew their policies with the TT Club each year.

The project to develop a VCM was part of the strategic planning cycle. The TT Club wanted to better understand their strategic value drivers, with an emphasis on the nonfinancial and intangible drivers of performance. The development of the map involved a set of interviews with members of the senior management team, the CEO, as well as the board members. In a facilitated 1-day planning workshop with the senior management team, the map was finalized. The VCM for the TT Club is outlined in Fig. 4.15.

The TT Club decided that their value proposition was to provide sustainable financial security for the global transport industry by offering excellent and customized insurance covers and value-added services that people trust. They identified three core competencies: (1) the claims handling and delivery of services such as risk assessments and advice; (2) the deep understanding of the industry and changing client demands and underwriting requirements; and (3) the ability to build and maintain close relationships with the Industry, which gives the TT Club the status of an independent body of the industry.

These competencies are delivered through the structures, processes and systems in place, together with the reputation and recognition of the TT Club as a specialist and member of the transport industry. These competencies are also delivered through relationships not just with the transport industry, but also with reinsurers and brokers. At the foundation of the VCM is the ability to recruit, train, develop and retain good people who help to create the knowledge and expertise needed. This knowledge, together with the strong customer-care ethos, helps to shape the TT Club's reputation in the industry. It also shapes the development of processes, structures and systems.

Another enabler at the heart of the strategy is capital strength and access to reinsurance, one of the strongest resources in the TT Club. There is a dynamic relationship between the relationships with reinsurers and the access to reinsurance. Capital strength is also an important driver of reputation; without appropriate capital strength, the reputation would soon suffer. The TT Club's global presence helps it to create local relationships, which in turn help its reputation and recognition in the field. The headquarters in London enable the TT Club to develop the crucial relationships with brokers who sell their products and with reinsurers to make reinsurance deals.

Case Study: The Motor Neurone Disease Association[32]

The Motor Neurone Disease (MND) Association is a charity dedicated to the support of people with MND and those who care for them. MND is a progressive neurodegenerative disease that attacks the upper and lower motor neurones. Degeneration of the motor neurones leads to weakness and wasting of muscles, causing increasing loss of mobility in the limbs, and difficulties with speech, swallowing and breathing. One famous person affected by this incurable disease is Cambridge Professor Stephen Hawking, who wrote the all-time best seller A *Brief History of Time*.

The MND Association has a vision of a world free of MND. The mission of this great charity is to fund and promote research to bring about an end to MND. Until then, it will do all that it can to enable everyone with MND to receive the best care, achieve the highest quality of life possible and die with dignity. The charity will also do all that it can to support the families and careers of people with MND. The Association was formed in 1979 by a group of volunteers who wanted to coordinate support, guidance and advice for people affected by the illness. It now has 3000 volunteers and 120-plus paid staff, all dedicated to improving the lives of people affected by MND, now and in the future.

The VCM for the MND Association was created in a series of workshops attended by their top management team and was supplemented by reviews of existing strategy documents, reports and analyses. Figure 4.16 shows the VCM of the MND Association. The overall output stakeholder value proposition is to play

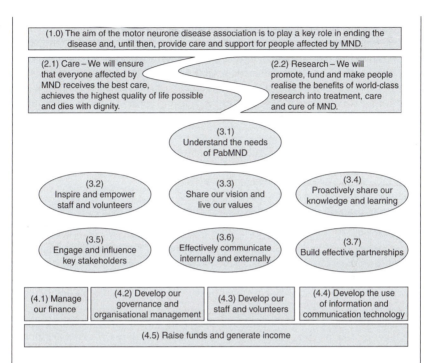

FIGURE 4.16   MND Association value creation map.

a key role in eradicating the disease and, until then, providing care and support for people affected by MND. The MND Association identified two core activities, namely, care and research. On the care side, the charity will ensure that everyone affected by MND receives the best care, achieves the highest quality of life possible and dies with dignity. On the research side, it will promote, fund and make people realize the benefits of world-class research into treatment, care and cure of MND.

In order to fulfill its value proposition and deliver its core activities, the MND Association identified the following value drivers and enablers:

● To understand the needs of people affected by MND
● To inspire and empower their staff and employees
● To share their vision and live their values
● To proactively share their knowledge and learning
● To engage and influence key stakeholders
● To effectively communicate internally and externally
● To build effective partnerships
● To manage their finance
● To develop their governance and organizational management
● To develop its staff and volunteers
● To develop the use of information and communication technology
● To raise funds and generate income to support their activities

Case Study: The National Lottery Commission

The National Lottery Commission is responsible for licensing and regulating the National Lottery of the United Kingdom. Over 70% of adults play lottery on a regular basis, and the crossed fingers logo is recognized by 95% of the UK population. National Lottery Games consist of draw-based games, such as Lotto, scratchcards and interactive instant win games. There are a number of ways in which people can play the National Lottery:

- At one of approximately 26 000 National Lottery retailers throughout the United Kingdom
- Over the Internet at www.national-lottery.co.uk
- Via FastPay outlets at supermarket checkouts
- Via interactive digital television
- Play by text

The money generated by the game is roughly broken down as follows: 50% is paid to winners in prizes, 28% is given to good causes; 12% goes to the government in lottery duty; 5% is paid to National Lottery retailers on all National Lottery tickets sold; and 5% is retained by the operator to meet costs and return to shareholders.

The role of the National Lottery Commission is to protect the integrity of the lottery, protect players, and maximize funds to good causes. The organization also runs the competition to grant license to the selected operator of the lottery. The National Lottery Commission is a nondepartmental public body, sponsored by the Department for Culture, Media and Sport. It operates at arm's length from government and its decisions are independent. Its work is funded by the National Lottery Distribution Fund (NLDF).

The Commission has the following powers:

- It runs the competition process to select the operator of the lottery.
- It makes sure that the operator meets the conditions of the license and it can take legal action if necessary.
- If the operator breaks the conditions of the license, it can impose a financial penalty.
- Ultimately, it could withdraw a license, but that would only happen only under extreme circumstances.

With the aim of being a world-class regulator, the National Lottery Commission embarked on a strategic performance management initiative and designed a VCM to clarify and agree their forward strategy. Based on interviews of directors and senior managers, a draft map was produced which was later refined in a series of workshops. Figure 4.17 shows the VCM of the National Lottery Commission. The output stakeholder value proposition is to regulate the lottery to ensure that players are treated fairly, the nation's interest in the lottery is protected and the operator is motivated to maximize the enjoyment and benefits that the lottery brings to the nation.

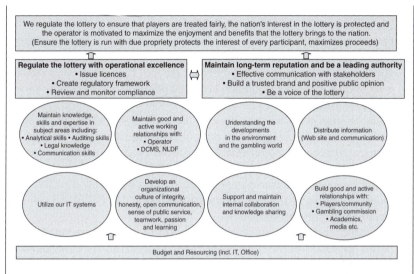

We regulate the lottery to ensure that players are treated fairly, the nation's interest in the lottery is protected and the operator is motivated to maximize the enjoyment and benefits that the lottery brings to the nation. (Ensure the lottery is run with due propriety protects the interest of every participant, maximizes proceeds)

Regulate the lottery with operational excellence
- Issue licences
- Create regulatory framework
- Review and monitor compliance

Maintain long-term reputation and be a leading authority
- Effective communication with stakeholders
- Build a trusted brand and positive public opinion
- Be a voice of the lottery

Maintain knowledge, skills and expertise in subject areas including:
- Analytical skills • Auditing skills
- Legal knowledge
- Communication skills

Maintain good and active working relationships with:
- Operator
- DCMS, NLDF

Understanding the developments in the environment and the gambling world

Distribute information (Web site and communication)

Utilize our IT systems

Develop an organizational culture of integrity, honesty, open communication, sense of public service, teamwork, passion and learning

Support and maintain internal collaboration and knowledge sharing

Build good and active relationships with:
- Players/community
- Gambling commission
- Academics, media etc.

Budget and Resourcing (incl. IT, Office)

FIGURE 4.17 National Lottery Commission.

Its two core activities are (1) to regulate the lottery with operational excellence and (2) to maintain the long-term reputation as a leading authority on lottery matters.

The following underpinning enablers and drivers of performance were identified:

● Maintaining a good and active working relationship with the lottery operator
● Continuing to develop our specialist knowledge, skills and expertise in subject areas such as analytics, auditing, legal and communications.
● Utilizing our IT systems such as our control systems and remote access systems
● Nurturing an organizational culture of integrity, teamwork, learning and honesty and in which people communicate openly, have a sense of public service and a passion for what they do.
● Fostering internal collaboration and the sharing of knowledge
● Continuing to understand the latest developments in our environment and the gambling world
● Providing and distributing information through our Web sites and other communication channels
● Building good and active relationships with the player community, gambling commission, media etc.

These actual case examples, from a very diverse set of organizations, illustrate the type of outputs that can be expected from the vital process of creating a VCM and accompanying VCN.

SUMMARY

- This chapter outlines how to bring the different aspects of your strategy together into one integrated visual strategy map.
- The two major benefits of strategy maps are that they (1) ensure a *strategy is integrated and consistent* and (2) enable powerful and easy *communication of the strategy.*
- *Value creation maps* are introduced as the latest tool to visualize an organization's forward strategy. They bring together the three key elements of strategy, namely, its *output stakeholder value proposition,* its *core activities,* and its enabling strategic elements or drivers of performance. These three components are then placed in relationships with each other and visualized on one piece of paper to create a completely integrated and coherent picture of the forward strategy.
- The *output stakeholder value proposition* (or *output deliverables*) is a one or two sentence statement that answers the question of why an organization exists and what its roles and deliverables are. It identifies the key output stakeholders of the organization and describes what value the organization is delivering to them. It is mainly derived from the analysis of the core purpose and the output stakeholder requirements.
- The *core activities* are the few vital things an organization has to excel at in order to deliver the above value proposition. The core activities essentially define what an organization has to focus on and what differentiates it from others. Core activities derive from the assessment of the core competencies but translate them into actionable activities on which the organization will focus in the next 12 months. An organization would usually have between two and four core activities.
- The *enabling strategic elements* (or *value drivers*) are the other strategic elements or objectives an organization has to have in place or has to deliver in order to perform its core activities and meet its output stakeholder value proposition. These enabling elements or value drivers derive from the assessment of the resource architecture as well as the assessment of the input stakeholders and represent activities linked to an organization's financial, physical and intangible resources. An organization would usually have between 5 and 15 value drivers.
- A VCM is a unique representation of an organization's strategy and has a limited life span (usually 12 months, or in line with the organizational planning cycle). Like the strategy, a VCM has to be revised and renewed to keep it relevant.
- Real case studies from leading government, public service and not-for-profit organizations have been provided to illustrate how they have applied the VCM approach in practice.
- A short and concise written description of the strategy should be produced. A so-called *Value Creation Narrative (VCN)* tells in 500–1000 words how

the organization intends to create value by specifying its value proposition, required core activities and key enabling resources.

- A clearly defined strategy forms the basis for good strategic performance management. In the remainder of Part I of the book, I discuss how to ensure it is aligned with budgeting processes, project and program management as well as risk management. In Part II, I look at how the VCM can guide the development of meaningful performance indicators.

REFERENCES AND ENDNOTES

1. Sun, Tzu (1981). *The Art of War*. Hodder & Stoughton, London.
2. This analogy is borrowed from Kaplan, R. S. and Norton, D. P. (2000). Having Trouble With Your Strategy? Then Map It. *Harvard Business Review*, Sept–Oct, 167–176.
3. Deutsch, G. and Springer, S. P. (1998). *Left Brain, Right Brain*. W. H. Freeman, New York; Gardner, H. (1996). *Leading Minds – An Anatomy of Leadership*. Basic Books, New York.
4. Gardner, H. (1996). *Leading Minds – An Anatomy of Leadership*. Basic Books, New York.
5. Bukh, P. N., Larsen, H. T. and Mouritsen, J. (2001). Intellectual Capital and the 'Capable Firm': Narrating, Visualising and Numbering for Managing Knowledge. *Accounting, Organizations and Society*, 26(7), 735–762.
6. For a good overview of how mapping is used in strategic management, see: Huff, A. S. and Jenkins, M. (2002). *Mapping Strategic Knowledge*. Sage, London.
7. For a detailed description of what a balanced scorecard is, see: Marr, B. (2008). *What is a Balanced Scorecard?* Management White Paper, The Advanced Performance Institute (www.ap-institute.com); Kaplan, R. S. and Norton, D. P. (2000b) (see note 2 above); or Kaplan, R. S. and Norton, D. P. (2004a). *Strategy Maps – Converting Intangible Assets into Tangible Outcomes*. Harvard Business School Press, Boston, MA; Kaplan, R. S. and Norton, D. P. (2000a). *The Strategy Focused Organization: How Balanced Scorecard Companies Thrive in the New Business Environment*. Harvard Business School Press, Boston, MA.
8. Kaplan, R. S. and Norton, D. P. (2000b) (see note 2 above); Kaplan, R. S. and Norton, D. P. (2004b). Measuring the Strategic Readiness of Intangible Assets. *Harvard Business Review*, 82(2), 52–63.
9. Marr, B. and Adams, C. (2004). The Balanced Scorecard and Intangible Assets: Similar Ideas, Unaligned Concepts. *Measuring Business Excellence*, 8(3), 18–27.
10. See for example: Atkinson, A. A., Waterhouse, J. H. and Well, R. B. (1997). A Stakeholder Approach to Strategic Performance Measurement. *Sloan Management Review*, Spring, 25–37; Maltz, A. C., Reilly, R. R. and Shenhar, A. J. (2003). Beyond the Balanced Scorecard: Refining the Search for Organizational Success Measures. *Long Range Planning*, 36(2), 187–204; Ahn, H. (2001). Applying the Balanced Scorecard Concept: An Experience Report. *Long Range Planning*, 34(4), 441–461; Marr, B. and Adams, C. (2004) (see note 9 above).
11. Marr, B. (2008) (see note 7 above).
12. See note 11 above; Neely, A., Adams, C. and Kennerley, M. (2002). *The Performance Prism: The Scorecard for Measuring and Managing Business Success*. FT Prentice Hall, London.
13. One of the earliest uses of influence diagrams was by J. Forrester to represent a causal loop in a feedback system. Later, Professor Ronald Howard from Stanford University and his colleague, Dr James Matheson, refined and popularized influence diagrams. See: Howard, R. A. and Matheson, J. E. (1990). *Principles and Applications of Decision Analysis, Volume I*. Strategic Decisions Group, Menlo Park, CA; Howard, R. A. and Matheson, J. E. (1990). *Principles and*

Applications of Decision Analysis, Volume II. Strategic Decisions Group, Menlo Park, CA; Howard, R. A. (1965). Dynamic Inference. *Journal of the Operations Research Society of America*, 13(5), Sept.–Oct., 712–733.

14. Initial strategy maps showed linkages between individual objectives [see Kaplan, R. S. and Norton, D. P. (1996b). Linking the Balanced Scorecard to Strategy. *California Management Review*, 39(1), 53–79], whereas later templates only visualize relationships between the perspectives [see Kaplan, R. S. and Norton, D. P. (2004a) in note 7 above.]

15. For examples of influence diagrams, see: Gupta, O. and Roos, G. (2001). Mergers and Acquisitions Through an Intellectual Capital Perspective. *Journal of Intellectual Capital*, 2(3), 297–309; Marr, B., Pike, S. and Roos, G. (2005). Strategic Management of Intangible Assets and Value Drivers in R&D Organizations. *R&D Management*, 35(2), 111–124.

16. See, for example, the work by John Sterman at the Systems Dynamics Research Group at MIT; Sterman, J. (2000). *Business Dynamics: Systems Thinking and Modelling for a Complex World*. McGraw-Hill.

17. See, for example, Marr, B. (2004). *Business Performance Management, Current State of the Art*. Cranfield School of Management and Hyperion Solutions, San Francisco (this report can be downloaded from www.ap-institute.com).

18. All these are real examples, some have been annonomized to protect the businesses. Being real life examples, they do not necessarily represent best practice but illustrate how organizations have translated the theory into practice to make it work for them.

19. For the full case study on how DHL applied the value creation map, see Marr, B. (2006). *Strategic Performance Management*. Butterworth-Heinemann, Oxford.

20. Professor Jan Mouritsen of Copenhagen Business School, with other colleagues from Denmark, proposed the use of narratives for providing contextual information for intellectual capital statements. See for example: (DATI) Danish Agency of Trade and Industry (2000). *A Guideline for Intellectual Capital Statements – A Key to Knowledge Management*. Ministry of Trade and Industry, Copenhagen.

21. Scott recommends the adoption of the language used in a serious conversation, meaning using only as many words as necessary. See: Scott, R. (1989). *Secrets of Successful Writing*. Reference Software International, San Francisco.

22. For more details see: www.shell.com

23. See for example: Becker, B. E., Huselid, M. A. and Ulrich, D. (2001). *The HR Scorecard: Linking People, Strategy, and Performance*. Harvard Business School Press, Boston, MA; Keyes, J. (2005). *Implementing the IT Balanced Scorecard: Aligning It With Corporate Strategy*. Auerbach Publishers, Philadelphia; Graeser, V., Pisanias, N. and Willcocks, L. P. (1998). *Developing the IT Scorecard: A Detailed Route Map to IT Evaluation and Performance Measurement Through the Investment Cycle*. Business Intelligence, London.

24. See, for example: Nicholson, N. (1997). Evolutionary Psychology: Toward a New View of Human Nature and Organizational Society. *Human Relations*, 50(9), 1053–1078.

25. This case study is based on: Marr, B., Cregan, R, Husbands, E. and Millar, G. (2007). *Measuring and Managing Performance in Local Government: Belfast City Council*. Management Case Study, The Advanced Performance Institute (www.ap-institute.com).

26. For more information, please see: Marr, B. (2006) (see note 19 above).

27. This case study is based on: Marr, B. and Shore, I. (2008). *Cascading Balanced Scorecards: Using Strategic Maps to make Performance Relevant to RAF Stations*. Management Case Study, The Advanced Performance Institute (www.ap-institute.com)

28. This case study was jointly produced with Emma de-la-Haye, Carol Jones and Mark Rigby from the Performance Management and Governance Team.

29. Please note that the relative importance and relationships between the resources has been taken out to protect confidentiality and competitive information.
30. This case study is based on: Marr, B., Watters, J and Weir, M. (2007). *Measuring and Managing Intangibles: The Scottish Intellectual Assets Centre*. Management Case Study, The Advanced Performance Institute (www.ap-institute.com).
31. This case study was jointly produced with Paul Neagle, CEO of the TT Club and Nick Baker, Business Planning Director at the TT Club.
32. For more information, please visit: http://www.mndassociation.org

Chapter 5

Aligning Your Organization with Your Strategy

You don't have to be a brain surgeon to understand that once you have agreed a strategy that your organization has to be aligned and managed appropriately to the delivery of this strategy. This means you need to ensure that the different activities, programs and projects help to deliver the strategy, that the budgets are allocated accordingly, that your organization is structured suitably and that the key strategic risks are managed. This sounds simple and straightforward, right?

Unfortunately, this level of alignment is extremely rare in government, public service and not-for-profit organizations. Why is this so? I believe one of the key reasons for this is the lack of strategic clarity. If organizations haven't got an agreed and clearly outlined strategy, then this presents a massive barrier to real alignment. However, once you have agreed on your strategy and visualized it in a value creation map, the process of aligning activities, budgets, structure and risk management is no longer complicated.

If your value creation map outlines the forward strategy for your organization and makes explicit what it is there to do, the core activities it needs to excel at, as well as the enablers of good performance, then all activities in your organization need to be aligned with this map of your strategy. The questions I will address in this chapter include:

- How do we align our activities, projects and programs with our strategy?
- How do we align our budgets and budgeting process with our strategy?
- How do we align our structure and governance processes with our strategy?
- How do we align our risk management activities with our strategy?

ALIGNING ACTIVITIES

In order to perform to the best level, everything an organization does, all its activities, need to be related to the delivery of its overall objectives. When I talk about activities here I refer to any tasks that are performed within an organization, but mainly to those that are brought together into projects and programs. I define a project as a set of planned, coordinated and collaborative delivered tasks to achieve specific organizational objectives. A program, on the other hand, is a system of projects that is coordinated to deliver specific organizational objectives.

A good starting point for assessing alignment between current activities and the strategy as outlined in the value creation map is to identify all major initiatives (projects and programs) that are currently being run or those that are planned for the future. This list can then be used to map the different activities and initiatives against the elements on the value creation map (see Fig. 5.1).

Such mapping exercises regularly reveal a number of activities that cannot be mapped against any of the strategic elements on the strategic map. This indicates that these projects do not directly contribute to the implementation of the new strategy. The implication of a mismatch between strategy and activities can be twofold. First, it can be used as a checking mechanism to see whether anything important has been missed out from the strategy map. If so, the strategy might have to be revised to include these. However, it is important not to see this as an excuse to just add new elements to the map for every activity or initiative that would not fit into the existing strategy. Adding a new element to the value creation map at this stage should be extremely rare. Second, and the much more likely implication, the organization is doing things it shouldn't. This means a serious discussion needs to take place about the reasons for doing projects that are not aligned with the current forward strategy and in most cases should lead to project closure.

Once you have mapped your existing and planned activities onto the value creation map and any gaps are sorted out by either adding elements to the strategic map or by eliminating unnecessary projects, it is now time to review how well this portfolio of activities is delivering the strategy. A new strategy wouldn't be new if nothing needs changing. The balance of current or planned

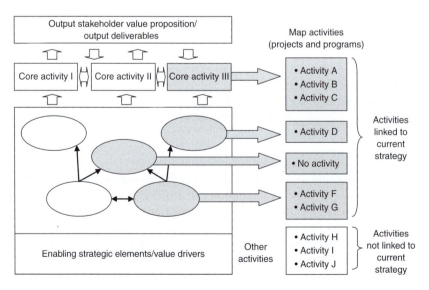

FIGURE 5.1   Mapping current and planned activities.

activities is obviously optimized for the previous strategy. As a consequence we often find that a few of the strategic elements have a lot of activities linked to them, whereas others have only few or no projects linked to them (also indicated in Fig. 5.1). The former tend to be objectives and activities that were in place before, whereas the latter tend to be new ones.

Especially in a world of limited resources, which most public service and not-for-profit organizations operate in, decisions have to be made about the new portfolio of activities. The elements on the map with no existing or planned activities obviously require the creation of new activities and initiatives. At this point it often dawns on people that things need to change if they are serious about their new strategy. If new activities have to be introduced, then it often means that other activities have to be stopped or cut back. A good starting point is to rank the activities for each element on the value creation map to identify which ones are the most important or essential activities to deliver on the different objectives and activities.

When we work with our clients, we usually create a so-called *heat map* to facilitate this decision-making process. A heat map is a color-coded value creation map indicating different levels of current performance for each of the elements on the value creation map. We use green to indicate good performance, yellow to indicate minor problems, orange to indicate some major problems and red to indicate serious underperformance. Such a heat map can be produced in workshops with the management team or in my case I often produce them based on the initial interview data collected to draft the value creation map. These interviews are usually so rich that they provide sufficient information to make reliable judgments about each of the elements on the map.

Not surprisingly, the elements with few or no activities tend to be red on the heat map and the elements with many existing initiatives tend to be green or amber. Where this is not the case and many initiatives still lead to poor performance levels, it is time to review the activities itself to understand whether they are the wrong activities or they are being badly executed.

Once the elements on the strategy map have been prioritized, performance levels have been assessed, and the associated activities have been ranked and prioritized within each of the strategic elements, a picture emerges about how well the current and planned activities are aligned with the strategy and what activities need to be introduced or eliminated to ensure a good delivery of the new forward strategy (see Fig. 5.2). My experience of working with many government, public service and not-for-profit organizations from all across the world has taught me that management teams in these organizations really don't like making these kinds of decisions and as a consequence sometimes shy away from making them. The harsh consequence of this, however, is that nothing will change and the new strategy will never be delivered unless the activities of the organization are closely aligned with the elements identified on the value creation map.

This emerging picture of the activities deemed necessary to deliver the strategy has to be translated into a business plan. The creation of a business

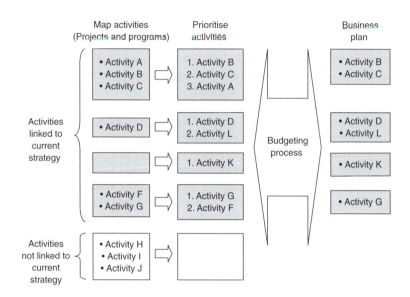

FIGURE 5.2   Prioritizing activities and deriving the business plan.

plan should be an iterative process that interacts with the budgeting process (also indicated in Fig. 5.2). Budget constraints often set a ceiling on funding and therefore introduce a sense check on the overall number of activities that the organization can realistically afford. Understanding the budget limitations allows organizations to decide on the final realistic set of activities. Many organizations including Belfast City Council have now replaced their traditional business planning process with the process outlined here where the strategy depicted in the value creation map provides the necessary guidance to create a business plan. In the next section, I will discuss the alignment with budgeting processes in more detail.

ALIGNING BUDGETS

It is important that you align the budgeting process with your organizational strategy. The value creation map captures this strategy and allows organizations to identify and prioritize activities (as outlined above). Once the strategic objectives and underpinning activities are agreed, a budget should be set to ensure the agreed activities (tasks, projects and programs) are funded. However, a budget also needs to be flexible enough to allow organizations to reallocate budgets at regular intervals as needs arise or performance levels change. While the value creation map might remain valid for 12 months, the underlying activities to deliver the strategic objectives tend to require more flexibility. As a consequence, organizations should be in a position to regularly reforecast required budget levels and reallocate resources.

However, the budgeting process in most government and not-for-profit organizations is characterized by an annual budgeting process that starts sometime in late summer or early autumn to set a fixed annual budget and financial targets for the following financial year. The annual budgeting round kicks off the budgeting game of forward and backward negotiations in which everybody tries to maximize their budgets while at the same time trying to commit to as little as possible. These are numerous iterations in which guesswork coupled with negotiating skills finally lead to an agreed budget. This budget is usually quite detailed and broken down into specific annual allocations of money for different tasks. Once the budget is agreed, people start playing the numbers game. We have all seen this where organizations become extremely creative about how to allocate their spending to meet their set budget. In addition, we can often see organizations in the government sector that go on a spending spree just before their budget period comes to an end to ensure they have spent all their money in order not to incur any potential cuts for the next budget round. Below I summarize some of the major and well-documented problems with the annual budgeting process.

- The process takes too long and is too resource intensive, which costs organizations huge amounts in money and time.
- Budgets are often created based on guesswork or 'sophisticated' tool such as '5% more than last year'.
- Organizations create a rigid funding framework that is often out of date by the time the new financial year starts.
- Organizations often end up running their company to their set budget instead of the ever-changing strategic priorities.

In his report titled 'Reinventing planning and budgeting for the adaptive enterprise', James Creelman outlines why, despite the widely acknowledged problems with the annual budgeting process, budget still remains common practice. He finds that the reasons for its continuing use include the fact that it is well understood, is institutionalized as an annual ritual, is seen as an effective control mechanism and clearly assigns accountability and its perceived value as an incentive or compensation mechanism.[1]

It is staggering that only just over half of organizations closely align their budgeting process with their strategic performance management process.[2] To me, it is just common sense to align your spending with your strategic objectives and priorities, and I believe that decoupling the strategic performance management process from the budgeting process is one of the biggest barriers to good strategic performance management. If you run a budgeting process that is separate to the strategic performance management activities, you run the danger of misalignment of funding and strategic objectives. What often happens is that the budgeting process 'wins' and people focus on meeting the budget rather than the strategic priorities.

There are different ways of moving beyond the traditional budgeting process.[3] What I recommend is to start with your business plan as outlined above and then use this to set an initial budget. However, instead of setting this as a rigid annual budget, I suggest that organizations produce rolling forecasts that allow them to take a more flexible approach to their budgeting to ensure they can alter their budgets to react to changes in their environment. One practical way to do this is to set aside a certain amount of money that is not allocated to 'business as usual' projects and activities but can be used throughout the year for new or changing activities.

Case Study: Scottish Enterprise

There are not many examples of public service and not-for-profit organizations that have been brave enough to step away from the rigid annual budgeting process, but Scottish Enterprise is an organization that has done this.

Scottish Enterprise is Scotland's main economic development agency, funded by the Scottish government. Its mission is to help the people and businesses of Scotland succeed. In doing so, they aim to help build a world-class economy in Scotland. Headquartered in Glasgow, it employees about 2500 people who provide following services:

- Helping new businesses get underway
- Supporting and developing existing businesses
- Helping people gain the knowledge and skills they will need for tomorrow's jobs
- Helping Scottish businesses develop a strong presence in the global economy, building Scotland's reputation as a great place to live, work and do business.

Scottish Enterprise was one of the first government organizations to replace their annual budgeting process with the basic performance management approach outlined in this book.

Scottish Enterprise uses their Balanced Scorecard to set clear strategic objectives and the management team became increasingly uncomfortable with the disconnect between the setting of strategic objectives and the annual budgeting process. Today, they have moved away from the traditional budgeting approach to a process that is based on strategic plans, which are supported by resources that are 'drawn down' for key projects as required rather than the traditional fixed allocation of budgets. Quarterly performance reviews lead to rolling forecasts and allow them to draw down more or less resources as required.[1]

ALIGNING ORGANIZATIONAL STRUCTURE AND GOVERNANCE

When a new strategy is being designed, it also makes sense to take a look at the way an organization is structured and governed. I have worked with many organizations where it was necessary to change the organizational structure or governance to create a better fit between the strategic priorities outlined in the value creation map.

If you have identified new strategic themes and objectives, then for me it is only common sense to align the organizational structure and governance to these themes. Many public sector and not-for-profit organizations are structured in very traditional and functional silos instead of cross-departmental structures that are aligned with the corporate objectives. Similar to unaligned activities, strongly hierarchical or siloed structures that are not closely aligned with the strategy can pose serious barriers to the strategy implementation. I find that especially in government organizations there can be a reluctance to embrace new ways of working, and they are often unable or unwilling to address the fragmentation and silos within their organizations.[4]

A good way to understand the alignment between the strategy and the current structure is to take a look at the value creation map and the associated activities and identify who within the organization, that is which groups of people or departments, will be responsible for delivering them. If it is easy to map the different departments and group to individual elements and objectives on the map then this indicates good alignment. However, if it is difficult to map individual groups or departments to the different tasks and if the different elements and objectives require collaboration between different subgroups of departments, then it might make sense to consider restructuring the organization and governance processes to better align it with the strategic objectives.

Some organizations have addressed this with a matrix structure that allows them to keep functional departments but overlay a cross-departmental structure designed to deliver the strategic objectives identified in the value creation map. By its simplest definition, the matrix is a grid-like organizational structure that allows an organization to have multiple command structures. In a basic two-dimensional structure, an organization would have the functional structure, but in addition it would also create a cross-functional structure to deliver the various strategic objectives identified on the value creation map.

This can be less disruptive to individuals as the functional or departmental structures will remain as a constant but the silos are broken up by the cross-departmental structures that are managed to focus on the strategic priorities. The advantage of this model is that it gives organizations the flexibility to address the changing strategic priorities without having to change all the lines of authority and responsibility. However, it also has some disadvantages that have to be managed carefully. Perhaps the key disadvantage is the creation of two lines of supervision, with its potential to create role conflict, role ambiguity and role overload. Good communication mechanisms need to be put in place to ensure the potential drawbacks are addressed. I will discuss the communication element later in this book chapter 9, in the section on performance reporting and performance reviews.

ALIGNING THE MANAGEMENT OF RISK

I strongly recommend that you align the management of your risks with the strategy outlined in your value creation map. Public service and not-for-profit organizations face many potential areas where they are vulnerable to significant risks, and it is important that these risk factors are actively managed. Risk management as a management tool started to emerge in the 1990s (although the problem has of course been around for much longer than that!). However, the main emphasis has been on financial risks and external risks. Financial risks are concerned only with financial uncertainties, whereas external risks are often identified in the external strategic analysis. More recently, organizations have started to look at risks more holistically to identify possible threats to their business model and value creation.

Many organizations admit that they do not have processes in place to effectively manage strategic and operational risks. One of the key problems has been for organizations to identify the areas where they face risks and are vulnerable. Because organizations still find it difficult to identify everything that matters, they often revert to the traditional areas of risk: financial and external. Too many organizations still treat risk management and performance management as separate functions instead of integrating them. It is important to understand that risk management and performance management do in fact represent two sides of the same coin – if the performance management approach identifies the key drivers of performance, then it is only natural that those are the areas for which firms require risk indicators too.

Your value creation map is a visual representation of your business model and all the components required to deliver your value proposition. It should, therefore, be used to also guide the risk assessment and allow organizations to identify potential focus areas for their risk mitigation strategies. The value creation map might therefore identify that government and not-for-profit organizations face risks related to core activities and value drivers such as corporate reputation, information and data security or staff motivation. Your organization's unique value creation map can be used to assess the risks in the value proposition as well as the risks concerning the core activities and enablers identified to deliver the value proposition. In this way, organizations can cover all areas they believe are important for their business and are able to weigh up the potential significance of the risks that they face.

Case Study: Royal Air Force[5]

The Royal Air Force (RAF) in the United Kingdom has been able to align their performance management with their management of risks. Once the strategic elements and objectives are identified, the next step is to link performance management with risk and issue management. Performance management is about measuring, reporting and making decisions about achieving objectives. Risk management is about identifying and managing the risks to achieving the same objectives. Thus, there is a common denominator – an organization's objectives.

For example, key objectives are expressed as the delivery of the Force Elements at Readiness (FE@R), which are reported as part of the performance management approach. For example, the performance management system might identify shortfalls between the targets that have been set (e.g. 10 crew members at 7 days readiness) and the achievement (e.g. 8 crew members at 7 days readiness). The shortfall of two crew members is defined as an issue if it is certain. If it is only probable, it is classified as a risk. Any such shortfall, which will be revealed by the performance management process, needs to be identified, assessed and managed as in all good risk management practices.

The performance management process within the RAF also includes the forecasting of the position in the future, extrapolating from the known baseline of the current position. For such forecasting to be realistic it needs to take into account the known risks and issues and modify the extrapolation to reflect the assessed most likely impact of the combined effect of these risks and issues. Thus, judgments being made within the performance management process cannot be separated from data available from the risk management process. They are inextricably linked.

The RAF has been doing risk management for many years. However, unlike a commercial company, the risk focus is not primarily on the financial 'bottom line' but rather on the delivery of war-fighting capability. Key risks for the RAF are those that impact on war-fighting capability, termed, loosely, 'operational risk'. Some of these are generated by deliberate decisions shaped by available funding. For example, to limit the level of spares purchased to avoid overspending a budget raises the risk that equipment will be unavailable due to unserviceability. Again providing a lower (cheaper) level of protection for personnel, risks increasing the danger to them in a hostile environment. Over time, these risks accumulate and their combined effects may be greater than those recognized when looking at the individual risks.

An IT system called SAPPHIRE was designed to meet commanders' need for greater visibility of their 'operational risks', and to maintain both an authoritative audit trail of reviews and the current state on ameliorating actions, all to aid their decision making. It also had to meet the corporate governance requirements; however, these were considered to be secondary to the prime objective.

SAPPHIRE has therefore been designed to provide both performance and risk information in a manner that allows the performance assessments to be made in the light of the risks that have been recorded. The risk and issue management implementation was an enduring success. After an initial 3-month pilot period on selected stations, the system was introduced command-wide. To avoid the important being swamped by unhelpful detail, commanders were encouraged to concentrate on their top 5–10 concerns.

ANALYZING RISK

The first step in assessing risk, therefore, must be to identify possible areas of risk. The best way to do this is to take the value creation map and go through all its elements, identifying potential risks (see Fig. 5.3). Below, I will highlight some common risk areas in the external environment together with examples for each of the resource and performance driver categories identified in Chapters 3 and 4.

External Risks

Many of the external risks will usually be identified and addressed in the external strategic analysis (see Chapter 2), when organizations look at the political, economic, social, technological, environmental or legal conditions in their sector and markets. Also, the five forces framework will identify threats from competitors, suppliers and so on. Here, I will briefly look at two common external risks – competition and market risks.

Competition Risks

Risks in this category can range from the emergence of a new supplier to the market to the threat of a competitor organization developing a superior product, service or process that is difficult to replicate, and which allows it to capture market share from other incumbent suppliers to that industry. Many government organizations do nowadays compete with private sector competitors

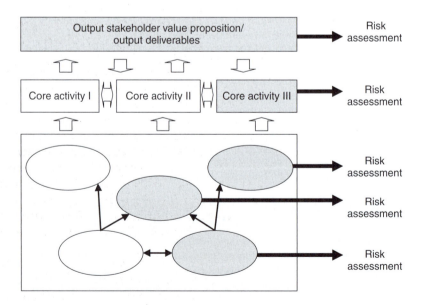

FIGURE 5.3   Identifying potential risk areas.

for customers and many not-for-profit organizations face competition similar to for-profit commercial enterprises.

Market Risks

This is simple Darwinian theory of adaptability to environment changes. If customers stop buying (or reduce their consumption of) products or services due to economic or other environmental factors, then this will generally affect all players in that marketplace, be it government organizations, not-for-profit enterprises or commercial companies. However, those who accurately identify the trend and react fastest to changes in market demand will normally be in the best position to survive a market downturn and prosper in the future, while others will be less fortunate. As Gary Hamel and C. K. Prahalad observe, 'The cues, weak signals, and trend lines that suggest how the future might be different are there for everyone to observe'.[6]

Financial Risks

Monetary or financial risk is an area with which many organizations are quite familiar, especially government and not-for-profit organizations, where funding is a precious resource that needs to be actively managed. Here, I will briefly look at cash flow or capital risk, which is one of the risks that is relevant for government and not-for-profit organizations. For financial risk management, organizations deploy practices to optimize the manner in which they take financial risk. This involves upholding relevant policies and procedures, such as monitoring the risk that the cash flow of an organization will be adequate to meet its financial obligations. Many companies use hedging as a technique to reduce or eliminate financial risk by, for example, taking two investment positions that will offset each other if prices change. A prominent example of when financial risk management strategies go wrong can be seen in the case of Barings Bank. One of its young traders, the now infamous Nick Leeson, went to Singapore and was trading a very low risk strategy of just betting on the same futures contracts in two different markets in Asia and basically just buying low and selling high. At first he was phenomenally successful and was regarded as a hero by Barings, which gave him more money to trade. Subsequently, when Nick lost a little bit of this money, he managed to cover up those losses by hiding them in a separate account. However, when trying to make back those losses, he began to take much bigger risks with much larger sums of money. This produced further losses when the markets went against his expectations, so that he needed to hide these too and, as we all know, this resulted in the collapse of the entire bank.

Risks Related to Physical Resources

There seems to be an increasing risk to our physical resources due to more frequent natural and man-made disasters. A series of natural disasters and increased levels of international terrorist activities have both contributed to a heightened awareness of these risks. However, these are not the only types of physical resource risks, some are much more mundane. Here, I will examine disaster risk and bottleneck risk.

Disaster Risk

Organizations and its stakeholders alike need the comfort of assurance that in the event of a major catastrophe, such as a devastating fire, bomb attack, airplane crash or a natural disaster at one of their vital premises, a close to normal service can be provided by the organization very rapidly after the event. For example, following the 9/11 attacks several financial services companies that used the World Trade Center and other nearby buildings were able to resume customer service activities within just a few hours from backup facilities they had set up for such a catastrophic event (though few would have predicted the severity of it). The same is true of London-based companies following the damage caused to the Baltic Exchange by IRA bombs in 1992. With the global threat from terrorism, many government organizations face an increasing disaster risk and have to put in place mitigation strategies.

Bottleneck Risk

If Eliyahu Goldratt taught manufacturing companies anything in his ground-breaking 1984 novel *The Goal*[7], it was the simple fact that a breakdown, failure or delay within a key constraint part of an organization (in this context, a particular machine tool or production section on the factory floor) creates a problem for the *whole* plant. There are areas of vulnerability in almost all organizations where, if glitches occur, the resulting impact will be far greater than if there is a failure elsewhere, which can relatively easily be recovered.

Note to all call center operators: make sure your telephone lines are always available and, in emergencies, offer a call-back facility that works – the enormity of the harm that bad call centers can do has become legendary, but so real too in terms of retaining customer loyalty. Internal capacity constraints need to be recognized and managed – they are 'arteries' that must be kept open at all costs. Although this has been classified here as a physical resource risk, and it generally is, the problem it defines can also be about key people in the organization too (see below) – particularly critical decision-makers and authorizers.

Risks Related to Human Resources

A key risk that is regularly overlooked in organizations is risk related to its staff and to the knowledge they possess. Organizations are often unaware that there might be some individuals with critical knowledge and expertise who could walk out any day. Another associated risk is the fact that knowledge is not only important, but also a very vulnerable resource – it tends to deplete over time if it is not nurtured. Furthermore, unfortunately, a small percentage of employees may not necessarily be as trustworthy as we would like them to be. Here, I will look at knowledge risk, staffing risk and employee theft risk.

Knowledge Risk

Like tangible assets, knowledge has to be maintained to retain its value. Knowledge that is not kept up-to-date can quickly lose its value or even disappear. Our civilization seems to have lost the knowledge of how the Egyptian pyramids were built; even with modern calculation techniques it cannot be explained how these structures remain standing. Knowledge, like all resources, is context specific. Changes in the external environment can make knowledge and skills redundant, as many craftsmen experienced during the industrialization process that took place in the nineteenth and twentieth centuries.

For example, companies that held a lot of knowledge and expertise about how to build a typewriter experienced how quickly this knowledge can become redundant with the arrival of the computer. Today, knowledge can be very short lived. Computer programs may be standard one day, but can be replaced by new innovative programs the next day. It is critical therefore for organizations to understand the value of their knowledge and ensure that continuous training keeps knowledge up-to-date.

Staffing Risk

The impact on productivity of disaffected staff not being engaged with their organization's objectives can be substantial. While strikes and increased levels of absenteeism provide evidence of extreme levels of employee dissatisfaction, more subtle disaffection is achieved by slowing down, not answering the telephone, being rude to customers, gaming imposed performance measures and so on. A number of studies have verified the positive link between satisfied employees and happy customers, particularly in retailing (e.g. the well-known Sears, Roebuck and Company case illustrates this[8]). It doesn't work on a 'stand-alone' basis though because the service element has to be good too, but it is a vital success factor nevertheless. Organizations need to carefully monitor the pulse of employee perceptions about the organization and their relationship with it. In addition to employee morale, organizations need to be aware of their staffing needs both in terms of numbers of employees and the skill-sets with which they need to be provided. The availability of authorized staff to make particular decisions is an important facet of this equation too.

Employee Theft Risk

Employee theft risk is the principal reason that most large organizations have an internal audit department. It is also the reason why many organizations appear to have elaborate control procedures that seem to exhibit a lack of trust in their staff. Some staff are dishonest, albeit a small percentage, and there are many examples of employees (often in collaboration with others) who have ripped their employers off for considerable sums of money. For example, in the United States, theft by staff in retail stores alone is estimated to have reached a level of $14.9 billion in 2000;[9] employee theft is responsible for more than 46% of what the retail industry calls 'shrinkage' – far more than theft by shoplifters.

Another study by KPMG of 5000 businesses, agencies and nonprofit organizations in 1998, revealed that losses *averaged* $624000 from check fraud by employees, including forgeries and mailroom theft; looting of company bank accounts by employees came in at an average of $300000 per organization; theft and misuse of company credit cards amounted to an average of more than $1.1 million; and the average loss from expense account abuse was $141000.

When the Association of Certified Fraud Examiners, based in Austin, Texas, conducted one of the first comprehensive studies of employee fraud in 1996 in the United States, it reported that organizations typically lost 6% of their annual revenues to such theft.[10] There is research that suggests organizations may be able to reduce the risk of theft by creating what I call an enabled learning environment (see Chapter 11).

Risks Related to Structural Resources

Risks to structural resources include threats to organizational processes and routines, especially those posed by losing database contents and software because of hackers and viruses. There is also an increasingly common risk of intellectual property theft as well as the danger to business success created by more powerful regulatory regimes that are rightly intolerant of 'old school' exploitation practices. Below, I will observe and illustrate each of these risks.

IT Systems Risk

Hackers, viruses, worms and the likes have created a whole new industry in computer protection. Many large not-for-profit and governmental organizations' computer systems have been paralyzed in the last few years by malignant individuals with high levels of knowledge about information technology protocols who are intent on exposing the vulnerabilities of IT systems and their contents. Apart from the damage they create, which has to be repaired, viruses cause lost data, lost work time and lost revenues (customers go elsewhere). While the Internet has many upsides, it does have a downside too.

Intellectual Property Rights Theft Risk

Luxury goods and technology companies are particularly, but by no means exclusively, prone to this type of risk. As I have noted earlier (in Chapter 3), intellectual property rights can take many forms: from data, trademarks, logos and characteristic styling to more mundane industrial patents and media copyrights. We have all seen shady traders in almost every major city and resort in the world that sell copies of Rolex watches, Gucci or Prada handbags, Louis Vuitton luggage, DKNY or Calvin Klein jeans and various fan-ware, such as New York Yankees caps or Manchester United shirts. And sometimes they are remarkably good copies. But, while you may think that this is relatively harmless because these traders are selling to a largely different set of consumers than the original brand owners do, the damage to the exclusivity of the brand is being done.

Many government and public sector organizations are starting to recognize the value of their intellectual property and are starting to sue the organizations that violate their rights. For example, organizations such as the National Geospatial-Intelligence Agency in the United States or the Ordinance Survey organization in the United Kingdom are both agencies that hold a lot of geospatial data and information that they own. The National Geospatial-Intelligence Agency is part of the US Department of Defense, which manages and provides imageries and geospatial information for diverse military, civil and international needs. The data can be useful for satellite navigation providers or online mapping providers; however, they need to license the usage of the data and can't just steel the intellectual property to it.

Regulatory Risk

The reason I have included regulatory risk within the structural resource category, rather than in external risks or stakeholder relationship risks – although glitches in this area tend to have an adverse effect on the latter too – is that the root cause is often a failure in the framing, communication or policing of internal policies. As the power of regulators has escalated in recent years, the risks of deliberate cheating and sloppy management are escalating too. Compliance issues are fundamental to doing business.

Today, organizations need to be sure that they are operating within regulatory rules and guidelines. The litany of major government and not-for-profit organizations that have been exposed and fined considerable sums for being involved in illegal activities expands almost daily in the press. For example, in 2005 alone, regulatory bodies in the United States and Europe have meted out substantial fines for accounting fraud, price fixing cartels, bid rigging, bribery, market abuse, mis-selling of financial services, mishandling of complaints, misleading advertising, failing to inform investors and sales of abusive tax avoidance schemes. Organizations need to examine where they are at risk from regulatory investigation and clean up their ethical acts.

Risks Related to Relational Resources

In today's networked world, relationships are crucial ingredients for all organizations in both the private and public sectors. Their reputation hangs on these vital relationships and often the risk needs to be cascaded through the supply chain that helps to deliver the products and/or services that the organization sells or provides. Here I will, therefore, examine reputation risk and supply chain risk.

Reputation Risk

Reputation risk is probably the most under-rated and least understood category of risk in this list. If an organization fails to live up to its declared (or expected) values and is then consequently exposed to adverse media attention, then the consequences can be catastrophic. This has the potential to instantly disenfranchise all of an organization's branding efforts. Naturally, this category not only includes both product quality failures (such as recalls and warranty claims) and customer service quality failures, but also negative media publicity. For example, Arthur Andersen, one of the 'Big 5' worldwide accountancy firms, imploded in just a period of weeks after a few of its partners were involved in several high-profile financial scandals that hit the worldwide media – its clients walked away in droves.

Since 1992, Nike has been the focus of international scrutiny of how huge Western companies treat their suppliers in some of the poorest parts of the world. It has frequently been accused of promoting the use of 'sweatshops' in Indonesia, Vietnam, China and South America, where labor abuses, forced overtime and unsanitary conditions abound. About 500000 workers in over 350 factories across the globe make Nike footwear and apparel. Activist groups, such as Global Exchange, first bombarded the media alleging abuses. By 1997–1998, an anti-Nike campaign led by human rights activists culminated in several 'Protest Nike' days in the United States. Many not-for-profit organizations face similar risks. Just think of charities that manufacture their own merchandise.

Nike's initial response was sluggish, but quickly gathered momentum when it realized the damage that could be done to its brand and also to its college campus sales. It, therefore, introduced a code of conduct for its suppliers, created a remediation plan and implemented independent monitoring of its suppliers' factories. It even published the location of many of these factories, which it had previously refused to do on competitive grounds. Lost reputation can not only disenfranchise customers, but also current and potential employees. The best people will move to the organizations with the best reputation.

Supply Chain Risk

Suppliers are a critical component of an organization's ability to deliver products and services to its customers, especially in the age of outsourcing 'non-core' activities. If a supplier defaults for reasons of capacity shortages, quality failures, a strike or a fire at their premises, then such disruptions are likely to have a major impact on customer service. Since the advent of 'Just-in-Time' delivery systems, which eliminate buffer stocks in the production system, such events can have a very rapid impact.

For example, when Ford's supplier of door and boot latches defaulted (not components that most people would associate with production criticality issues), plants in Dagenham and Cologne came to a halt. Consequently, output of nearly 3000 cars a day was lost and more than 10 000 workers were either sent home or diverted to plant maintenance.

In 2005, British Airways' sole supplier of in-flight meals at London's Heathrow Airport, Gate Gourmet, sacked over 650 unionized workers when a festering industrial dispute erupted into an illegal strike. (Gate Gourmet is part of a US private equity-owned company that was once part of British Airways (BA) before BA decided that catering was not one of its core competencies.) The situation rapidly escalated when 1000 BA workers at Heathrow, many of whom had relatives affected by the Gate Gourmet strike, started their own unofficial sympathy strike. This action forced the cancellation of flights and more than 100 000 passengers were stranded, mostly at Heathrow, one of the busiest airports in the world, causing scenes of anger and chaos at the height of the August holiday season.

As a corollary to this drama, it was revealed that BA had squeezed its supplier so hard in pricing negotiations that Gate Gourmet had no alternative but to reduce its mainly Asian-origin workforce substantially because the contract had become financially unviable (although, arguably, it could have examined other ways of parting company with them – to avoid insolvency of its UK operations).

This is a classic example of a clash of interests and cultures within the supply chain that affected all the key stakeholders adversely. The customer (BA), the supplier, the customer's passengers, both the supplier's and customer's employees, and the shareholders of both BA and Gate Gourmet were all losers. The trade unions involved may yet, at the time of writing, also suffer recriminations.

In recent years, during which high-profile cases of corporate wrongdoing have caught the attention of the media and NGO activists have become more vocal in attacking corporate behavior, reputation management has climbed the ladder of boardroom priorities. Reputations take years to develop, but can be destroyed very rapidly. Because of the Internet, the speed with which a reputation can be attacked by a broad range of different stakeholders – customers, employees, former employees, former suppliers, labor and human rights activists, and so on – on a global scale has been reduced to a matter of hours.

A 2003 survey[11] found that 60% of the world's CEOs view corporate reputation as a 'much more important' aspect of business than 5 years ago. The study also found that maintaining a good reputation has become so important that 65% of the world's CEOs have taken full responsibility for managing this aspect of performance. This figure rises to 80% in the United States, whereas in Europe it stood at just 44% (possibly due to closer relations between the CEO and the board). CEOs acknowledge customers as the external force with greatest effect on reputation, followed by print media, financial analysts and shareholders.

Risk assessment then is a highly significant factor for managing in today's business environment. So, organizations need to get to grips with the various risks they face. Given that there are many potential risks, it is advisable

TABLE 5.1 Risk Log

Key elements (from the value creation map)	Description of risk	Risk level: potential consequences/ Impact	Likelihood level/ Changing probability: probability that this risk will occur, changes in likelihood	Risk score: risk level + likelihood level	Accountability/ Review frequency: who is accountable, How often is this risk reviewed?
Employee knowledge	Our knowledge in Y software might become redundant if X becomes new standard	About a third of our leading programmers could become redundant	Not very high; most research shows that Y will stay the main standard; constant likelihood	3 + 1 = 4	Amanda Simon (quarterly)
Intellectual property right	Our patented software is copied in India and China	The copying could lead to significant revenue losses and loss of reputation	Very high – first reports indicate that this is happening; increasing likelihood	4 + 5 = 9	Peter Smith (monthly)
IT infrastructure					
Reputation					
Others					

to begin accumulating data that gives organizations useful information about where they are most exposed. In the next section, I will discuss how the value creation map can be used to analyze and evaluate potential risks.

These risks can then be captured in what I call a 'risk log'. This is a table that can be used to capture, describe, assess and quantify potential risks (see Table 5.1). This often requires obtaining factual information about these risks and then prioritizing their relative importance. Organizations need to assess the potential risk areas for their component parts, categorize them and then assess which are most important to manage.

COMPLETING A RISK LOG

In a risk log, organizations can capture their key risks. It can become a working document that is part of the performance management system. Below, I outline the various steps involved in creating such a risk log.

1. For each element on the value creation map, potential risks are identified. This element-by-element approach ensures that all potential risk areas are discussed – both external and internal. Moreover, using the value creation map also helps organizations to identify how potential risk areas might impact each other. However, it is unlikely that all potential risks for each element are identified and prioritized straightaway. The risk log will usually grow over time as more potential risk areas are identified, but the relevance of some will also tend to fall away as they are either mitigated or become less relevant over time (see below).   -

2. Describe the essence of the particular potential risks for each element. Here, it is possible to give the risk a name, but more importantly to create a short narrative *description* of the type of risk.

3. Define the *risk level*. Here, the likely consequences and potential impact of this risk are evaluated for the case that the risk becomes a reality.

4. Define the *likelihood level*. Here, the likelihood that this risk might turn into a reality is evaluated. In addition, this likelihood is compared to the likelihood of the last review cycle. This indicates whether the likelihood is increasing, staying the same or decreasing.[12]

5. Ascribe an appropriate scoring system according to (a) the *risk level* (potential severity) of each risk (e.g. 1–5), the criteria for which may not necessarily be all financial ones and (b) the *likelihood level* (probability of occurrence) of the risk (e.g. 1–5). These two scores can then be added up to create the risk score. The rationale for this scoring system is not only to help identify management priorities, but also to assess whether the likely severity of each risk has moved over time and whether the firm's potential exposure to it has increased or diminished since the last review.

6. Assign responsibility (ownership) for managing each defined risk and define a review frequency for reevaluation of subsequent risk mitigation activities.

Completing the risk log is best done within a project team. Different sub-teams can be assigned to assess the risks of the different elements of the value creation map. This ensures that several people who are knowledgeable in the subject matter work together and come either to a unanimous or aggregate score. Here, teamwork is important because this type of analysis can be highly subjective. That being the case, it is a good idea to ensure that the risk level and likelihood scores are not left to a single individual. Furthermore, it is important to document as much information and logic as possible for the awarded scores in the risk log so that these can be revisited at the next review.

For each area, additional data can be collected and referenced in the risk log. However, there is a real danger here too of making this an overly bureaucratic process, and that is why I advocate a relatively simplistic approach. The Pareto principle applies: 80% of the risk can be identified and assessed with 20% of the potential effort required to do it.

Having identified the highest priority risks (with high risk level and high likelihood level), management actions can be taken to modify their consequences and potential impacts on the firm. Typical actions resulting from a risk analysis include:

- development of mitigation plans (especially for emergencies/crises – scenario planning techniques can assist this process),
- buying insurance against occurrence,
- renegotiation of supplier contracts,
- introduction of (internal/external) compliance audits,
- introduction of new performance indicators to monitor emerging trends.

This does not mean that lower severity/likelihood risks can be ignored altogether; it is just that management is unlikely to be able to set in motion the corrective actions for large numbers of risks simultaneously. However, if this is treated as part of an organization-wide program, then actions on lower priority risks might – with appropriate guidance – be delegated to lower ranking managers. Otherwise, they will have to wait until the senior executives have first dealt with the highest priority risk category and that might mean that the firm is still exposed to some pretty substantial risks with which it is unready to cope.

Although organizations have certainly been at risk for many centuries (how else would the insurance industry have become so wealthy?), arguably they have never been *so* at risk. Today, it is becoming increasingly common and necessary for organizations to appoint a senior risk manager. This is a post that often reports to a nonexecutive director but where the incumbent needs to work closely with operational executives in far-flung parts of the organization. Introducing an evaluation methodology that is aligned with the corporate strategy and then conducting a fair assessment of the potential risks is the first step toward mitigating the likely impacts that those key risks could have on the organization.

SUMMARY

- In this chapter, I have outlined the importance of aligning your organization with your agreed and mapped strategy in order to make the strategy real and ensure it gets implemented.
- *Activities need to be aligned* with the strategic objectives outlined in the value creation map. Activities are all tasks, projects and programs performed within an organization. A project is a set of planned, coordinated and collaborative delivered tasks to achieve specific organizational objectives. A program is a system of projects that is coordinated.
- The value creation map can be used to *map current and planned activities* in order to assess the match between the new strategy and the activities planned. The mapping of activities often reveals a mismatch between the activities and the forward strategy, which might require stopping some activities and starting others.
- A so-called *heat map* can be used to *prioritize* the strategic elements. A heat map is a color-coded value creation map that indicates different levels of current performance for each of the elements on the value creation map. In addition, projects and activities for each strategic objective can then be ranked to identify the most important activities to achieve each of the strategic objectives.
- The prioritized activities can then be translated into the operational business plan. To finalize any business plan, it is important to take into account any budgetary constraints. Therefore, the budget also needs to be aligned with the strategy outlined in the value creation map.
- Many government and not-for-profit organizations are still using *outdated annual budgeting processes*. The problem is that the process tends to take too long and is too resource intensive. Budgets are often created based on guesswork or 'sophisticated' tools such as '5% more than last year', which lead to a rigid funding framework that is often out of date by the time the new financial year starts. As a consequence, organizations often end up running their company to their set budget instead of the ever-changing strategic priorities.
- I have outlined a more flexible approach closely aligned with your strategic forward plan that allows organizations to *regularly reforecast and reallocate resources* to meet changing needs.
- Organizations need to align their risk management practices with their strategy. I suggest that *risk management is the flip side of performance management* which means risk and performance management have to be integrated.
- Because the value creation map is a visual representation of your business model and all the components required to deliver your value proposition, it should be used to also guide the risk assessment and allow organizations to identify potential focus areas for their risk mitigation strategies. I have outlined how to align risk and performance management and how to design a risk log for each of the strategic elements on your value creation map.

REFERENCES AND ENDNOTES

1. For more information please see: Creelman, J. (2006). Reinventing Planning and Budgeting for the Adaptive Enterprise: Tools and Techniques for Reengineering the Budgeting Process. Optima Publishing, London

2. For an in-depth discussion and the full research, please see: Marr, B. (2004). Business Performance Management: Current State of the Art. Research Report, Cranfield School of Management and Hyperion, San Francisco, CA, available to download from: www.ap-institute.com

3. For more information on moving away from the traditional budgeting process, see for example: Home, J. and Fraser, R. (2003). Beyond Budgeting: How Managers Can Break Free from the Annual Performance Trap. Harvard Business School Press, Boston

4. See for example: Achieving Innovation in Central Government Organizations, The National Audit Office, London, 2006

5. The case study is based on: Marr, B. and Shore, I. (2007). Measuring and Managing Performance in the Royal Air Force, Management Case Study. The Advanced Performance Institute, Milton Keynes, UK. (available from the resources section of the API website: www.ap-institute.com)

6. Hamel, G. and Prahalad, C. K. (1994). *Competing for the Future*. Harvard Business School Press, Boston, MA.

7. Goldratt, E. (1984). *The Goal*. North River Press, New York

8. See for example: Kirn, S. P., Quinn, R. T. and Rucci, A. J. (1998). The Employee-Customer-Profit Chain at Sears. Harvard Business Review, 76(1), 83–97; Heskett, J. L., Jones, T. O. and Loveman, G. W. (1994). Putting the Service-Profit Chain to Work. Harvard Business Review, 72(2), 164; Heskett, J. L., Sasser, W. E. and Schlesinger, L. A. (2003). The Value Profit Chain: Treat Employees Like Customers and Customers Like Employees. Free Press, New York, NY, p. 250

9. National Retail Security Survey

10. Winter, G. (2000). Taking at the Office Reaches New Heights. *New York Times*, July, 12

11. Conducted by communications consultancy Hill & Knowlton and executive headhunters Korn/Ferry (Reported in the Financial Times, October 14, 2003)

12. See for example: Drzik, J. and Slywotzky, A. J. (2005). Countering the Biggest Risk of All. Harvard Business Review, 83(4), 78–88

Part II

Collecting the Right Management Information

Once you have agreed, defined and mapped your strategy, you can use measurement to track progress and gain relevant insights to help manage and improve performance. Measurement has a central role in our society and organizations. Indicators help us to make sense of the world around us, allow us to put things in perspective and help us to make better-informed decisions.

We are used to applying the principles of measurement routinely in most aspects of our daily lives. When we watch the weather report on the television, we learn about expected temperatures, hours of sunshine, quantity of rainfall or wind speed. When we drive our cars, we are used to checking the speed or fuel consumptions; when we play or watch our favorite sports, we are eager to keep score and measure lap times, number of goals, home runs and shots on and off target; when we follow our favorite recipe, we measure the amount of flour or butter; and when we go to our doctors, they measure our blood pressure, cholesterol levels, heart rate or body mass index. All of these measurements serve to reduce complex elements of our lives to indicators in order to make them more digestible for us.

Measurement is deeply ingrained in everything we do. In school, we are used to test scores; we rely on clocks to get to work in time; we use food labels to compare calories, salt and fat content; and we use opinion pools and surveys to gauge attitudes and public perceptions. Traditionally, measurement was viewed as valid only if a numerical value could be obtained, and we often make the assumption that any measure has to be expressed in numbers.[1] Take this quote:

> ... when you can measure what you are talking about and express it in numbers you know something about it; but when you cannot measure it, when you cannot express it in numbers, your knowledge is of a meager and unsatis-factory kind.

This passing comment by Sir William Thomson (later Lord Kelvin), in a lecture to the Institution of Civil Engineers in 1883, is one of the most frequently cited quotes in meas-urement circles. The sentiment of this statement can be traced back to the philosopher Philolaus, in the fifth cen-tury BC, who said that 'without numbers, we could under-stand nothing and know nothing'.[2]

However, measurement in our modern world goes beyond numbers and includes using words to describe and assess performance. Measurement goes beyond the assign-ment of numerals and is much more of a social activity. Just think about choosing a restaurant for the next spe-cial occasion. You reflect on your previous experiences of the restaurants you have visited and might read reviews of new restaurants on restaurant websites or restaurant guide books in order to form an opinion about the different restaurants in your area. Based on the different reviews, ratings and your previous experiences, you then subconsciously, or even consciously, rank different elements such as food quality, service, atmosphere and price to choose the right restaurant for this occasion.

When we talk about performance measurement in organizations, we don't have to accept the limitation that any measure has to be expressed in a numerical form. Words, numbers, star ratings or traffic lights are all valid forms of measurement. What matters the most is that you measure the relevant things that will help you answer the questions that matter the most in your organization.

My experience and research of the Advanced Performance Institute shows that instead of identifying what they want to know and then designing the most meaningful performance indicators to help them gain the required management information, the majority of public sector and not-for-profit organizations just measure everything that is easy to measure but not what actually matters the most. The practice I have observed in many public sector organizations is something that I have 'scientifically' termed the ICE approach, which goes as follows:

- Identify everything that is easy to measure and count.
- Collect and report the data on everything that is easy to measure and count.
- End up scratching your head thinking: What the heck are we going to do with all this performance data stuff?

In this second part of the book, I will look at performance measurement more closely and provide easy-to-follow tools and techniques to ensure that relevant and meaningful management information is collected that helps organizations with their decision making and performance improvement. In Chapter 6, I first look at the different reasons and challenges for measuring and collecting performance information in public sector and not-for-profit organizations, before I introduce the concept of key performance questions (KPQs)[3] in Chapter 7. I believe KPQs are one of the most important recent innovations in strategic performance management and a key to successful strategic performance management implementations. KPQs are formulated before any indicators are designed to ensure that every indicator will help you answer a management question that actually matters. In Chapter 8, I outline how, based on your KPQs, you can design solid, relevant and meaningful performance indicators. See also Fig. P2.1 for an overview of Part II of this book.

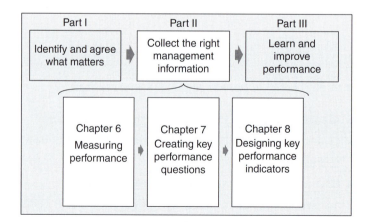

FIGURE P2.1   Collecting the right information.

REFERENCES AND ENDNOTES

1. Stevens, S. S. (1968). Measurement, Statistics, and Schemapiric View. *Science*, 161, 849–856; Stevens, S. S. (1951). Mathematics, Measurement, and Psychophysics. In: *Handbook of Experimental Psychology*. Wiley, New York.
2. Quoted in Boyle, D. (2001). The Sum of Our Discontent: Why Numbers Make Us Irrational. Texere, New York.
3. The concept of key performance questions was developed by Bernard Marr and the terms 'key performance question' and KPQ are trademarks of the Advanced Performance Institute.

Chapter 6

Measuring Performance

In this chapter, I explore the role of measurement in public sector and not-for-profit organizations and try to answer why we need performance indicators in those organizations. I discuss the challenges of measurement in a political context and the importance of measurement to facilitate learning and improvement. I also take a look at the limitations of measurement and its implications for the usage of performance indicators. The questions I address in this chapter include:

- Why do we need performance measures?
- What is the role of measurement in government, public sector and not-for-profit organizations?
- What can we really measure, and where are the limitations of measurement?
- What implications do the limitations have for the usage of performance measures?
- How do we define performance measurement in organizations?
- What are the differences between measurement and assessment, and between a measure and an indicator?
- What are some key rules that will make performance assessments work?

As mentioned before, measurement plays an essential role in our society, and as human beings, we have an intrinsic need to measure what is going on around us. We measure all day long. For example, we use our senses to measure: We use our eyes to see, ears to hear and fingers to feel. This allows us to assess whether the things around us are big or small, far away or close by, loud or quiet, hot or cold, etc. We then take all this 'measurement information' in and interpret it to make sense of the world around us. Measurement information therefore allows us to understand the world we operate in, without which we would be completely lost and stumble around in the dark.

Measurement arose from our human need for knowledge acquisition and social interaction; man is in fact a measurer of all things.[1] In today's world, measurement facilitates trade and commerce and provides the foundation for science and progress. Measurements are vital ingredients of everyday life, sense making and human understanding. Examples of these ingredients include clocks, calendars, rulers, clothes sizes, heights, weights, floor areas, cooking recipes, sell-by dates, alcohol content, match scores, ring sizes, diamond gradings, pint markings, calorie counters, bank accounts, speedometers, interest rates, thermometers, rainfall gauges, medical examinations, body mass indexes and questionnaires, to name just a few.[2]

While we generally accept that measurement is necessary for our civilization to flourish, in a public sector and not-for-profit organizational context, we often feel that it reduces human and social complexities to inhuman or meaningless numbers.[3] It seems as if we are obsessed with the process of quantifying, counting and calculating abstract numbers with the aim of generating ever-growing and ever more objective data sets. Instead of being a meaningful way of helping us make sense of the world around us and to guide our organizational decision making, most performance measurement activities in public sector and not-for-profit organizations are a complete waste of everybody's time. Whole teams and departments with new job titles such as 'performance manager' and 'performance analyst' have been created – individuals who shed blood, sweat and tears to put performance measurement systems in place. Unfortunately, the results of these efforts are often simply an increased administrative measurement burden, very rarely producing new management insight, learning or performance improvement.

Even worse, as a consequence of poor performance measurement practices in our organizations, we see counterproductive and even dysfunctional behaviors where we do the 'wrong' things just to meet performance measures and targets. For example, I see hospitals trying to manipulate waiting time targets and sometimes even rejecting patients who need urgent treatment, just to meet targets. I see teachers who spend the majority of class time 'training' pupils to pass exams instead of inspiring them to learn and police forces arresting innocent people to keep up their performance statistics. The truth is that organizations are full of measurement used for self-aggrandizement, self-promotion and self-protection; measurement used to justify pet projects or to maintain the status quo; and measurement used to prove, rather than improve.[4]

Why is this happening? Why is measurement in our organizations so mechanistic, process driven and number focused? Why is it driving insane behaviors and is the source of so much frustration and discontent? Why do we naturally, routinely and comfortably use measurement in our daily life to guide our decision making? And why don't we generally end up obsessively collecting measurement data when we get on with our daily routines? Why do we also tend to be able to make much better judgments about what is an appropriate set of measures when it comes to our daily life, and why are we unable to understand when we have collected a sufficient amount of data to make decisions in our organizations?

The answer to these questions lies in the different ways we use performance indicators in our organizations. In order to provide a better understanding of this problem and some of the important implications for using measures in organizations, I differentiate three reasons for measuring performance in organizations, and in the following sections, I discuss the problems and limitations that come with them.

UNDERSTANDING WHY WE MEASURE

There are many different reasons why we measure performance in organizations. These are often reduced to simple homilies, such as 'you can't manage anything unless you measure it' or 'what gets measured gets done'. I believe all different reasons for measuring performance can be categorized under one of the following three headings (see also Fig. 6.1):

1. **Controlling behavior:** Measures are used in a top-down command-and-control fashion to guide and control people's behaviors and actions. Measures are used to set goals or rules, to objectively access the achievement of these goals and to provide feedback on any unwanted variance between achievements and goals. Here, the aim of measurement is to eliminate variance and improve conformity. In this context, measures are often tightly linked to reward and recognition structures.[5]
2. **External reporting and compliance:** Measures are used to inform external stakeholders and to comply with external reporting regulations and information requests. When measuring for external reporting and compliance purposes, any reports and associated indicators can either be produced on a compulsory basis, such as annual financial statements and accounts, or be on a voluntary basis, such as environmental impact reports, for example.
3. **Learning and empowerment:** Measures are used to empower employees and to equip them with the information they need to learn and make decisions that lead to improvements. In this context, measures are used as the evidence base to inform management decisions, to challenge strategic assumptions and for continuous learning and improvement.[6]

It is the third reason – measurement for learning and empowerment – that is the most natural way of using performance indicators and that will lead to the biggest performance improvements. This is what we as human beings do day

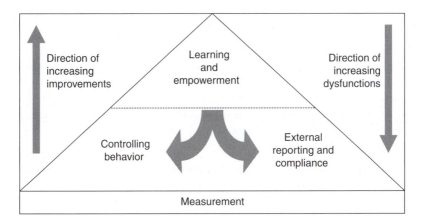

FIGURE 6.1 Reasons for measuring performance.

in, day out: We collect information to make sense of the world around us and use that information to guide our decision making and learning. However, the first two reasons for measurement often get in the way of real performance improvements. The reasons why measurement for controlling behavior and for external reporting and compliance get in the way of performance improvement can be traced back to the limitations of measurement. I discuss these limitations in the next section, before I come back to the three reasons of measurement and discuss the implications of the measurement limitations for the different ways we can and should use performance measurement data in our organizations.

WHAT WE CAN AND CAN'T 'MEASURE'

Measurement has been defined as the assignment of numerals to represent properties.[3] It is seen as the assignment of particular mathematical characteristics to conceptual entities in such a way as to permit an unambiguous mathematical description of every situation involving the entity and the arrangements of all occurrences of it in a quasi-serial order.[4] Whereas such technical definitions have been especially useful in disciplines such as physics, in the management field we need to rethink what we mean by performance measures.

Often the emphasis in measurement is on quantifications and numbers, with the intention to provide us with an *objective*, *uniform* and *rigorous* picture of reality. However, this seems to work better in some areas than in others. We find it easy to quantify things like money spent, the number of patients treated and the number of arrests made, and we can count incoming complaints, service visits or the number of refuse bins collected. Some things though are not easily counted. Things like overall service delivery, organizational culture, our know-how, the strengths of customer relationships or the reputation of your organization are all inherently difficult to simply count. At the same time, as we have seen from Chapters 2 to 3, the most important performance outcomes and enablers of future performance in government, public sector and not-for-profit organizations tend to be intangible in nature and therefore challenging to simply count.

Albert Einstein, one of the great thinkers of the twentieth century, emphasized that 'not everything that can be counted counts, and not everything that counts can be counted'. The problem arises when we try to use simple numbers to holistically 'measure' things that can never be measured completely or comprehensively. This is illustrated well by author David Boyle in his book on counting and numbers when he writes:

> We admit that numbers can't reveal everything, but we try to force them to anyway. We tend to solve the problem by measuring ever-more ephemeral aspects of life, constantly bumping up against the central paradox of the whole problem, which is that the most important things are just not measurable. The difficulty comes because they can *almost* be counted. And often we believe we have to try just so that we can get a handle on the problem. And so it is that politicians can't measure poverty, so they measure the number

of people on welfare. Or they can't measure intelligence, so they measure exam results or IQ. Doctors measure blood cells rather than health, and people all over the world measure money rather than success. They might sometimes imply almost the same thing, but often they have little more than a habitual connection with one another. They tend to go together, that's all.[7]

When it comes to many aspects that matter the most in our public sector, government and not-for-profit organizations, we have to rely on proxies or indirect measures,[8] which often capture only a fraction of what we want to measure (see, e.g. Fig. 6.2). The things we want to measure often have many different dimensions to them, but we regularly measure only one or two of these dimensions and, very often, are not even able to comprehensively measure each of those dimensions.

Take the measurement of human intelligence as an example to illustrate this point. We often make the assumption that an IQ (intelligence quotient) test 'measures' intelligence. However, does an IQ test measure intelligence? The answer is no. It focuses only on our analytical and mathematical reasoning. Dr Howard Gardner, professor of education at Harvard University, has shown that there are multiple dimensions to our intelligence[7] (similar to the different dimensions of the cube in Fig. 6.2).

Professor Gardner's studies have identified eight different dimensions of intelligence – of which an IQ test primarily addresses only one. The dimensions or forms of intelligence identified so far by Gardner are logical–mathematical intelligence ('number/reasoning smart'), as well as linguistic intelligence ('word smart'), interpersonal intelligence ('people smart'), bodily kinesthetic intelligence ('body smart'), spatial intelligence ('picture smart'), musical intelligence ('music smart'), naturalist intelligence ('nature smart') and intrapersonal intelligence ('self smart').

The first point therefore is that an IQ test focuses only on one out of eight possible forms of intelligence. This means that someone can be classed as intelligent when he or she, for example, has great hand–eye coordination and awareness of space – and therefore becomes a great basketball star or football

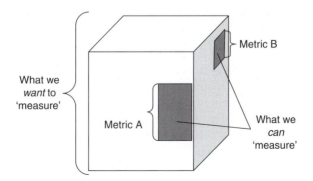

FIGURE 6.2   What we can and can't measure.

player. Someone can have great emotional intelligence and therefore be able to connect with other people and become a great leader.[8] Others might have great musical ability and become composers or musicians. All of these people wouldn't necessary need an exceptionally high score on an IQ test since it primarily assesses our logical and mathematical skills.

The second point is that an IQ test is still an imperfect measure of our logical and mathematical skills. One reason for this is that we can all train to pass IQ tests. We can read all the books on the topic, sit endless mock exams, get used to the type of questions, start to predict the answers they are looking for and therefore perform better and score higher. This means that even on the one dimension of human intelligence an IQ test is measuring, it is not a perfect or comprehensive measure of logical and mathematical skills. This is illustrated in Fig. 6.2 by the small gray square that represents what a metric can actually measure of the one dimension of intelligence.

From my experience, public sector, government and not-for-profit organizations are often prepared to sacrifice rich realities in order to achieve alleged rigor and clarity through oversimplified measures. This is why we use average Accident and Emergency Department waiting times as a measure of overall hospital service delivery and the number of police officers on the street and the number of arrests made as indicators for crime fighting. However, we have to be careful how we use such indicators. The American social theorist Daniel Yankelovich so rightly said that:[9]

- The first step is to measure whatever can be easily measured. This is OK as far as it goes.
- The second step is to disregard what can't be measured or give it an arbitrary quantitative value. This is artificial and misleading.
- The third step is to presume that what can't be measured easily isn't very important. This is blindness.
- The fourth step is to say that what can't easily be measured doesn't really exist. This is suicide.

Hans de Bruijn, professor at Delft University's Department of Public Policy and Management, also emphasizes this point in his book on public sector performance management when he argues that even though concepts such as comprehensive, consistent, clear or unambiguous have a strong 'feel-good' factor when used in relation to performance measurement, they don't really exist. He adds that if meaning is constructed based exclusively on the figures from a performance measurement system, then the meaning will be at best poor and probably wrong. Performance measurement is not fitted with dials from which performance can be readily read but rather with 'tin openers' that invite further investigation and interpretation.[10]

The above discussion hopefully illustrates that, in most cases, measures cannot capture the entire truth in an objective and comprehensive way. However, they can indicate a level of performance. They are therefore *indicators*, rather then *measures*, and have to be treated as such. What is important, however, is that we start acknowledging this fact and accept the limitations that come with it. Let me now discuss the implications of these measurement limits for the usage of performance indicators in our organizations.

IMPLICATIONS FOR THE USAGE OF MEASURES IN OUR ORGANIZATIONS

We have just established the fact that performance indicators cannot perfectly and comprehensively measure the things that tend to matter the most in today's organizations. However, being able to comprehensively and holistically measure performance is a pre-requite when using measures to control behavior or to ensure compliance. If we want to use measures to control people's behavior and to evaluate compliance, then we need objectivity and comprehensiveness. The reason for this is that, in both cases, objective indicators replace personal trust.[11] Here, measures provide a moral distance and make what is measured impersonal in a quest for objectivity. Objectivity is therefore required for the first two measurement reasons outlined above:

1. Using measures as a means of controlling people's behavior necessitates objectivity, especially if measures are tightly linked to reward and recognition. While in many public sector, government and not-for-profit organizations, the links to individual bonuses are used to a lesser extent, and reward and recognition does not necessarily mean financial rewards; it can mean simple praise, career progression or any other form of general recognition.
2. Reporting and compliance requires objectivity and, in many cases, even external auditing. Organizations use external auditors to provide an objective verification of the numbers they put into their annual reports. Some organizations go even further and also use external auditors to audit their numbers on voluntary reports, for example, environmental and social performance.

In both scenarios, personal trust is replaced with what need to be objective numbers. There is, in fact, a complex relationship between trust and quantification. For example, when farmers and merchants didn't trust each other to provide the right amount of wheat, they could use the standard barrel stuck to the wall of the town hall, which would measure the agreed local bushel.[12] It has been demonstrated that throughout history we were often able to win greater trust for claims by giving them quantitative expressions.[13] Nevertheless, it is dangerous to replace trust with measures since the big assumption is that we can measure everything that matters. The fact is that what matters the most in modern-day organizations is difficult to measure and impossible to express in objective and comprehensive numbers.

The problem is that if we believe our measures are objective and comprehensive and cover all important dimensions of performance, then this usually leaves blind spots in the measurement system that can be exploited and can lead to dys-functional behaviors. Let's go back to the cube in Fig. 6.2 and assume that the entire cube represents local government service delivery with all its facets. If we only measure whether the rubbish is being collected and whether the parks are free of junk, then people could just ensure that these dimensions are delivered, which can mean other dimensions of service delivery are neglected because they are not measured and as a consequence overall service delivery suffers. Such dysfunctional consequences of measurements can be seen in myriad examples. A good example is food standards. Like the farmers and the merchants using the standard barrel, today we rely on standards to facilitate international trade. The US food standards, which are administered by the Department of Agriculture, specify, for example, that a 'US Fancy broccoli stalk' has a diameter of not less than 2½ inches or that the color of a Grade A canned tomato is at least 90% red.[14] The same applies to the European Union, which specifies the standard bend of a banana or the size and shape of apples. We presumably all agree that what really matters are the intangible factors such as the taste and the nutritional quality of the produce, but these are again difficult to objectively measure. The standards are almost entirely based on the easy-to-measure physical appearance of the produce. And, in fact, studies have found that this has encouraged farm-ers to use dangerous pesticides not to increase yields but for the sole purpose of maintaining cosmetic appearance to meet such standards.[15]

In the next sections, I explore what this now means for using performance indicators in our government, public sector and not-for-profit organizations.

WHY MEASUREMENT FOR CONTROLLING BEHAVIOR ISN'T WORKING ANYMORE

Measures have, for a long time, been used to influence and control what people do. The theoretical model behind this is called 'agency theory'.[16] Its argument is that employees (agents) don't have the same objectives as the owner or insti-gator of an organization (principal). This is why the principal puts measures in place, links these measures to the reward system of the agent, which then guides the behavior of the agent and therefore aligns the objectives of the prin-cipal and the agent.

The widespread use of performance measures to control people's behav-ior can be traced back to scientific management defined by Frederick Taylor.[17] Taylor, one of the earliest management scientists, believed in using meas-urement to achieve conformity to rules. He held the view that there was one single best way to fulfill a particular task. According to him, it is only a mat-ter of matching people to a task and then supervising, rewarding and punish-ing them in accordance with their performance. In Taylor's view, there was no such thing as skill and all work could be analyzed step-by-step as a series of unskilled operations that could then be combined into any kind of job. Performance measurement systems used in the early years of industrialization reflected this view with a focus on operational efficiency and reward systems based on mechanistic measurement models. The stopwatch was the critical tool of that time.

There are some very good reasons why this model does not work anymore in modern organizations. First, such command-and-control models can work only if we can measure all critical dimensions of performance objectively and comprehensively. Second, there is an assumption in the model that we are only motivated by extrinsic rewards, that is, the payment we receive for completing our tasks. In today's world, and in government, public sector and not-for-profit organizations in particular, people are not only motivated by external financial rewards but also motivated intrinsically by doing good. This significantly weakens the underlying model and calls its applicability into question.

However, the biggest problem is caused by the blind spots in our measurement systems caused by our inability to measure everything that matters. During Taylor's time, the models worked well. When the first factories for mass production were being established, employees performed basic mechanistic tasks that could be measured very accurately. And if we only want people to come to work, for example, to stick screws into a piece of metal that is passing by on a conveyor belt, then we can measure this accurately and precisely and can reward them accordingly. The danger is that if the model does not measure some aspects of performance that are still important for the success of overall performance, then it leaves a gap or blind spot. And as we have seen from the discussion above, this gap between what we want to measure and what we can measure is endemic in modern organizations where people are required to do more than just perform simple mechanical tasks.

Marshall Meyer, professor at the Wharton School of the University of Pennsylvania, argues in his book on performance measurement that people will exploit the gap between what we want to measure and what we can measure by delivering exactly what is measured rather than the performance that is sought but cannot be measured.[18]

This causes dysfunctional behavior and suboptimal performance. Let's take the example of a public service call center, whose main objective is to answer service queries from members of the public. Measurement systems in call centers tend to focus on efficiency measures and cost-related measures with the hope that these would somehow also lead to better customer service delivery. However, it has long been established that it is dangerous to measure one thing while hoping to achieve another.[16] Typical measures used in call centers are the waiting time until a call is answered, the average call duration or call-handling time and the number of calls taken by individual call center agents – all of which are automatically produced by the system and therefore easily measured and reported.[19] The result is that because they are measured and reported, people assume that they are the most important and relevant metrics for the organization. This is all okay until we find out that frontline agents continuously transfer customers or even cut them off to meet their individual

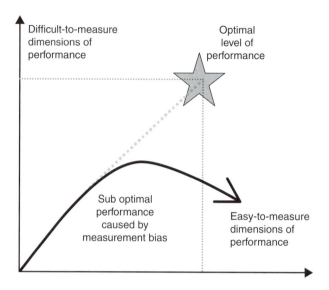

FIGURE 6.3   Biased command-and-control measurement systems.

call-handling time targets. While the efficiency targets are being achieved, customer service, and with it overall performance, is not. Figure 6.3 shows that optimal performance comprises some easy-to-measure dimensions, for example, average waiting time, average call-handling time and number of calls taken, as well as some more difficult-to-measure dimensions of performance, for example, quality of the call, first-time fix rate, empathy and end-to-end service delivery.[20] What usually happens is that only the easy-to-measure dimensions of performance are measured and then linked to a reward system (explicitly or implicitly). It is therefore not surprising that, in many call centers, frontline agents worry about and deliver only the easy-to-measure dimensions of performance such as the number of calls taken and average call-handling times. The fact that these calls might be of low quality and/or of little value to the customers is not taken into account (see Fig. 6.3).

Another good example comes from the UK government targets on waiting times to see your doctor. Since doctors are assessed on meeting a 48-h target from making an appointment to seeing the patient, most surgeries have now introduced a rule that patients can book appointments only up to 2 days in advance (even if they don't want one until next week). Therefore, just by changing their processes, every doctor meets his or her targets by default. This now adds another inconvenience to patients who have to ring back next week, but apparently, customer service has been improved if you look only at the numbers.

The creation of an environment in which trust is replaced with numbers to increase control causes social stratification. It is argued that imposing control measures on people will invariably activate the self-centered drives of organization members, and as a result, rank, territoriality, possessiveness, fear and anger will dominate social relationships. It has long been argued that no measurement system can be designed to preclude dysfunctional behaviors.[21] There are still people out there who believe that they can create a 'cheat-proof' measurement system by just establishing better and all-encompassing indicators. Theoretically, it is possible to close the blind spots in our measurement systems by measuring all dimensions of performance and by measuring them more comprehensively. However, in practice, it is not feasible because of the costs and efforts required to do this. Also, management Professor Meyer argues that compensating people for performance on multiple measures is extremely difficult. Paying people on a single measure creates enough dysfunctions. Paying them on many measures creates even more. The problem is combining dissimilar measures into an overall evaluation of performance and hence compensation. If measures are combined formulaically, people will game the formula. If measures are combined subjectively, people will not understand the connection between measured performance and their compensation.[22]

The above arguments therefore seriously question the usage of measures for influencing behavior in a command-and-control fashion. Controlling machines makes sense; however, trying to control the behavior of people creates negative and unpredictable consequences. For instance, showing a child how to cross the street by holding his or her hand for the first few times makes a lot of sense. Holding the child's hand for the rest of his or her life makes no sense at all and interferes with growth.[23] In many government, public sector and not-for-profit organizations, we confuse the need for accountability with the application of control. In fact, accountability and autonomy go hand in hand. While we need improved accountability, increased control through measurement will lead to a negative response to measurement and most likely dysfunctions.

I agree wholeheartedly with management Professor Charles Ehin, who writes that over the past 100 years or so, we have deliberately chosen to design our social institutions with almost one single purpose in mind – to control the behavior of people within them. However, he continues, success in the knowledge age demands that we let go of the top-down, command-and-control framework.[24] This argument is shared by many, including Robert Austin, professor at Harvard Business School, who also makes a very strong case that measurement for controlling people's behavior does not work anymore in today's organizations. Instead, we should focus our efforts on what he calls 'informational measurement' used for learning and strategic decision-making.[25]

WHY WE HAVE TO BE CAREFUL WITH MEASURING FOR EXTERNAL REPORTING AND COMPLIANCE

Measurement for external reporting and compliance has the aim of providing external stakeholders with information about the overall effectiveness, impact and/or outcomes of organizational performance.[25] It can be used to share

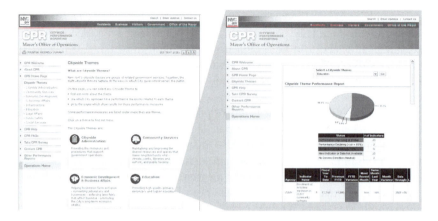

FIGURE 6.4   Citywide Performance Reporting (CPR) in New York City.

performance information with customers and stakeholders, to identify and share best (and worst) practice and to ensure compliance. Here, measurement can, for example, be linked to service performance reports for the public, statutory financial reports or external performance reports as well as audits by regulators such as the government audit offices or audit commissions.

There has been an increasing call for transparency and accountability in relation to efficiency and effectiveness of service delivery in government, public sector and not-for-profit organizations. As a consequence, more performance information is made available externally. This information provision either can be on a voluntary basis or can be prescribed by government rules or legislations. In addition, there are increasing auditing activities in government, public sector and not-for-profit organizations by which regulators or auditors either check the provided information or perform their own independent evaluations of performance. All of these reporting and audit activities increase the measurement burden for organizations. Let's look at each of these external usages of measures in a little more detail.

The honest and open reporting of measurement data and performance information to inform external stakeholders is a good idea and contributes to better communication and understanding between e.g. citizens and government organizations. An example is the Citywide Performance Reporting (CPR) initiative in New York City (see Fig. 6.4). The mayor of New York City said that they were proud to provide the public with the next-generation performance reporting, which makes performance across the city as transparent as possible. A Web site has been created to provide members of the public with citywide performance information at a glance.[26] From this Web site, users can view the critical performance indicators of over 40 agencies in the city, including the Fire Department of New York, Department of City Planning, Department of

Sanitation, City University of New York and New York City Police Department. Furthermore, visitors can review performance data for eight individual city-wide themes, which provide a performance view across different agencies and services. These themes include administration, education, infrastructure, legal affairs, public safety and social services. Besides the individual performance measures, the Web site also provides comparisons over time to indicate trends. Another similar example is provided by the Commonwealth of Virginia, which has developed a similar Web site.[27] Such reporting initiatives enable the public to monitor agency performance. The aim is to provide the public with relevant information on the most important 'outcome' measures – those that directly reflect how citizens' lives are affected by the government.[28]

While it is important to provide the public or external stakeholders with access to better performance information to improve transparency, which allows them to hold government, public sector and not-for-profit organizations account-able for their performance, the problem starts when this data is used by some to 'spin' the messages. The press or politicians have been known to misrepresent data, or show a biased view, for example, use only one or two indicators that indicate underperformance in order to support political attacks. I believe that to a certain extent this will not be avoidable in a political and social context and to some extent journalists, politicians and the public are encouraged to challenge performance data and question it. However, the fear of political spin often has as a consequence that organizations report only on very simple and objective measures that are difficult to challenge. Another implication is that organiza-tions tend to focus on simple output and process measures rather than measures of outcomes and performance enablers. One step into the right direction is to provide more interpretations of the performance data. Indicators need to be put into context, and it is important to explain what the performance measures actu-ally mean. If we provide stacks of raw data only, then we shouldn't be surprised to find that people use the data out of context or put their own interpretations or 'spins' on the information provided. This might provide some food for thought for reporting initiatives like the ones in New York City and Virginia, which cur-rently focus mainly on raw data.

Reporting performance data externally also allows organizations to bench-mark and compare themselves with other organizations. One of the benefits of this is that the better-performing organizations are then able to identify and share good practices with the poor-performing organizations to ensure that they are able to improve. However, there is a fine line between honest reporting and creating competition. Once we create competition between organizations, they are much less likely to share good practices. In fact, performance meas-urement becomes a disincentive for cooperation as organizations optimize their own performance and stop sharing any good practice. In some cases, they even hinder others in their improvement activities in order to stay ahead of their 'competitors'. This element of competition is created much quicker than many realize. For example, once we start producing publicly available league tables, the line is crossed and organizations are much more interested in measurement to 'look good' rather than to 'be good'.[29] A good way forward to avoid the element of competition is to avoid league tables and instead publish best practice case studies and performance turnaround stories that focus on learning without the element of obvious performance comparisons and competition.

Slightly worse is the problem when we have to report externally imposed performance indicators. They are often seen as a way in which government or external regulators impose their priorities on organizations without any consideration of local needs. And if there is a large number of externally imposed indicators and targets, then they can take over local performance measurement activities. Take for example the United Kingdom, where local government departments now have to report on about 200 performance targets, even if half of them are not relevant or important for some local governments. Officers in charge of performance measurement often spend the majority of their time collecting and reporting these externally imposed indicators so that there is little time to collect any other more meaningful performance information. As a consequence, these 200 measures are the only measures people in the organization see day in, day out and become the local performance measurement system. We therefore shouldn't be too surprised if people in our organizations don't think measurement is adding any value and is producing any relevant information. I believe it is important to separate externally imposed indicators that are not relevant to local strategies from the local measurement system. I am not suggesting ignoring government targets and regulator-imposed indicators; what I am saying is that we need to make sure every indicator is relevant to what matters in our organization. If you end up having to collect 200 indicators of which 100 are not relevant, then we need to keep them separate and tell people that we collect them only because we have to, in order to fulfill legislative reporting requirements. What we don't want to do is to dump all 200 indicators into the local measurement system and, by doing so, inflate and weaken the measurement system.

The auditing function is another important element of measurement for external reporting and compliance. Over the recent decades, an increasing number of audit and evaluation mechanisms have been introduced for government organizations, public sector bodies, charities and other not-for-profit organizations. In many countries, an entire industry has been created to conduct audits and some even talk about an audit explosion.[30] Auditing is seen as a mechanism to objectively and independently checking and verifying whether an organization has delivered its objectives. As a matter of fact, we as human beings constantly check up on each other and routinely monitor any stream of communicative exchanges that make up our daily life.[31] In his book on auditing, Professor Michael Power argues that checking and verifying is part of normal human exchange, which is mainly performed unconsciously. We therefore don't really think of it as 'auditing'. However, if we are aware of it and consciously perform an 'auditing' process, it tends to be in situations of doubt, conflict, mistrust and danger. In such situations, we check restaurant bills, make sure our kids are wearing seat belts, watch replays to see whether a ball was over the line, seek a second medical opinion, ask an expert witness, etc. Power writes:

> Methods of checking and verification are diverse, sometimes perverse, sometimes burdensome, and always costly. We normally reserve these actions for extreme cases. While it costs little to check that my children have fastened their seat belts in the car, the use of a private detective to check up on a lover can lead to obsession, despair, and even financial ruin, regardless of whether doubts and suspicions are verified. And there are circumstances where checking and demands for proof are not appropriate.[12]

Performance audits have their place; however, it all comes down to the way we use these audits. Some approaches are more useful than others. I suggest an auditing approach based on actual performance and with an emphasis on feedback and learning. If an audit report is positive and finds that an organization is performing well, then audits should be conducted less frequently and organizations need to be trusted to do a good job. If auditors identify elements of underperformance, the emphasis should be in providing learning, feedback and best practice examples to close the performance gaps. Unfortunately, this is not always the way these audits are conducted and many countries have created an army of auditors to audit every aspect of performance. In some cases this goes as far as auditing the auditing process, which clearly is a step too far and most likely a tremendous waste of everyone's time and money. I have just returned from a trip to Scotland, where a government body actually counted the number of audits it was subject to in a single year. The Highland Council established that within one year, 482 audits were conducted in the organization by 29 different regulators. They also calculated that this took up about 20% of management time. This shocking statistics is just an ordinary public sector example that illustrates the size of the problem. Few of these 482 audits were conducted intelligently, based on performance. In many cases, they were conducted in predetermined intervals (e.g. every 6 months) no matter how the council performed.

The fact that there are so many different ways of using measurement when it comes to external reporting and compliance can cause big problems. Let's take a local government organization as an example: This organization receives a number of government outcome targets and measures based on national priorities. Let's say there are 100 targets and measures. It now has to report on these 100 targets and indicators in every quarter. Then there are many local priorities, targets and measures about things like resource allocation, staff motivation and morale and service delivery. And on top of all this, there are a number of different external audits that require the collection of different data sets and measurements.

Let's imagine you are now a middle manager in one of the services units of this local government organization. All you see is a continuous stream of requests for measurement data for the quarterly national indicator set, for different local initiatives, and to satisfy the external audits. Of the 100 govern-ment targets, maybe only 60 are relevant to your organization and its local pri-orities, while the remaining 40 are not important but still have to be reported to generate national benchmarks. This makes it hard for you to see which indica-tors are actually important to your organization at this point in time and all you see is a hugely inflated measurement mess.

Another problem is that you are unsure about how the data will be used. The lack of clarity as to how measurement data will be used can create confusion and suspicion. You are constantly trying to assess whether the information you are providing will be used to judge your performance or check compliance, whether admitting underperformance might have resource allocation implications in the future, whether it will be used to identify the best practice or the biggest risks, whether it will be published externally or internally, etc.

In order to avoid this unnecessary inflation of measurement activities and the confusion and suspicion that come with it, we need to apply better filters to our performance measures and be a lot clearer about the purpose of each indicator we are collecting. There is a clear requirement for performance measurement to enable organizations to report to external stakeholders or comply with regulations and laws. This is not a problem as long as organizations understand that these measures are designed for that purpose only. Much care should be taken before any of those measures are used for any of the other measurement purposes. There is in fact a fine line between using measures of compliance and using measures for additional top-down control. If we cross the line to control, then we are back to the problem discussed in the previous section. Reporting performance informa-tion externally can also facilitate learning, which I believe is the key reason for measurement and discuss in more detail in the following section.

WHY MEASUREMENT FOR LEARNING AND EMPOWERMENT IS THE ONLY WAY FORWARD

I strongly believe that the only way forward is to focus our measurement efforts on learning. We have to use performance measures to gain insights that help us make better-informed decisions and improve performance. If we don't learn anything from the management information we are collecting and if it doesn't help us improve anything, then it is just a tremendous waste of every-one's time and efforts.

A key to this is to engage everybody in the organization in measurement activities. If measurement is seen as something the managers are using to check up on everybody or something that focuses on the reporting of abstract and seemingly irrelevant data sets, then it will be difficult to engage anyone. In order to achieve engagement, we have to use indicators to empower everyone in the organization. Everybody should be measuring performance and collect-ing his or her own performance data – just as we do in our private day-to-day life. We have to bring common sense back to performance measurement and ensure that we are collecting data that helps us answer the important performance questions in our organizations.

Using measures for learning means putting people in control of measurement. We need to not only get people to collect their own performance data but also get them to use the data to inform their decision making. And this will happen only if they see assessment as something that is valuable and something that is helping them to do a better job. The challenge for managers is to become more trusting. Managers have to concentrate on communicating direction and providing help to employees.[32] This thinking fits well with recent movements in government, public sector and not-for-profit organizations toward greater decentralization of management authority, autonomization of organizational subunits and enhancement of accountability.[33]

In this learning context, the meaning of measurement is broadened not to focus only on the narrow sense of measurement used in physics or mathematics. I prefer to use the words 'performance assessment', rather than 'measurement'. The process of assessment is therefore not about mechanistic quantification or the oversimplified assignment of numerals only. Performance assessment is about the systematic collection of information to enable everybody in the organization to evaluate performance and gain insights and learning.

In order to make performance measurement for learning and empowerment work, it is important to start with your value creation map and then identify the unanswered questions in relation to each of the objectives on your map. The so-called key performance questions (KPQs) allow you to identify your information needs and then guide you to more meaningful and relevant performance indicators.

When do we know we are getting there? I believe that when you get people in the organization asking for their performance data, when you get people who challenge existing data sets and indicators and when people start experimenting with their own new indicators, then this is a good indication that you are on the right track. In the following sections and chapters of this book, I focus on measurement for learning and empowerment and provide a number of practical tools to support this. Before I move on to introduce the concept of KPQs in Chapter 7 and discuss the design of meaningful and relevant performance indicators in Chapter 8, I define what I mean by performance measurement and recap some of the dos and don'ts of performance measurement in government, public sector and not-for-profit organizations.

SO WHAT DO WE MEAN BY PERFORMANCE MEASUREMENT?

We have learned that in any organization it is hard, if not impossible, to capture the whole truth in a small number of performance measures. I therefore prefer to use the word 'indicator', rather than 'measure'. A *performance indicator* 'indicates' a level of performance, but it does not claim to 'measure' it. If, for example, we introduce a new indicator to assess customer satisfaction levels, this indicator will give us an indication of how customers feel; however,

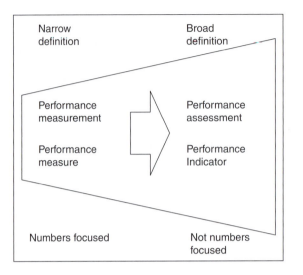

FIGURE 6.5   Towards assessments and indicators.

it will never 'measure' customer satisfaction in its totality. In the same way, an IQ test will give us an indication of intelligence but will never completely measure all dimensions of human intelligence.

I also prefer to talk about *performance assessment*, rather than 'performance measurement'. Performance assessment is a broader activity that takes into account not only numerals but also other forms of evidence such as written descriptions, observations, symbols and color codes. (see Fig. 6.5). Performance assessment goes beyond the technical aspects of collecting data and creating tables of numbers or scorecards.[34] Performance assessment is about using performance indicators to gain understanding and insights. It is about empowering people in the organization to make better-informed decisions that lead to improved organizational performance (see Fig. 6.6).

Technically, performance indicators identify to what degree a variable is present. Professor D. Lynn Kelly emphasizes that there is no reference to counting or quantifying the variable in such a definition.[35] The activity of performance assessment is the process of using the performance indicators to reduce uncertainty and to compare a given situation or status relative to known objectives or goals.[36] Performance assessment enables and empowers everybody within an organization to evaluate performance and gain insights that lead to better decision making and performance improvement.

In organizations, performance assessment should be clearly linked to the strategy, that is, the things that matter the most. Before any performance indicators are designed, KPQs are developed to identify the information needs. They identify the unanswered questions that the indicators help to answer. Let's now recap some of the rules that will make performance assessment work.

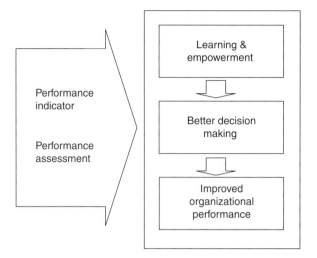

FIGURE 6.6   Defining performance assessment.

THE DOS AND DON'TS OF PERFORMANCE ASSESSMENT

In this section, I bring together some of the key dos and don'ts of performance assessment. All of them are critical success factors in any serious attempt to make performance assessments and performance management work in your organization.

- Do be clear about the functions of measurement and the way the indicators are used: Organizations often fail to be specific and clear about the function of the indicators and the forums for which they will be used. If this is the case, and the purpose is unclear, then it can create not only confusion but also suspicion about the use of performance information. Once a function and forum has been defined, no single user group is allowed to change the function without consulting the other users involved.[10]
- Don't use indicators for additional top-down control: Using measures for top-down control by management or by the government through its target regimes will result in gaming, that is, the reactive subversion such as 'hitting the target but missing the point' or ignoring areas of performance that are not measured or where no targets apply.[37]
- Don't tightly link indicators to incentives: Linking indicators to incentives is dangerous because they provide strong motivation for gaming and cheating: The higher the perceived incentive, the more perverse the effects. Professor Hans de Bruijn calls this the law of decreasing effectiveness.[38] It is important to keep in mind that incentives are not just financial rewards but also better reputation, which can come from better positions in public league tables.

- Don't just measure everything that is easy to count: We need to stop counting only what is easy to count because this creates countless problems. Management writer Charles Handy said, 'Measuring more is easy, measuring better is hard.' Too many organizations just brainstorm what they could possibly measure and then end up with a shockingly long list of everything that is easy to measure. Instead, we need to start with identifying what matters the most.
- Do link your indicators to your strategic objectives: Indicators need to be tightly and directly linked to the strategic objectives of an organization. This way we ensure that indicators provide information on what is most important for the organization. As the value creation map tells us the most important components of our strategy, it can also guide us to measure what matters.
- Don't let the government or regulators determine your measurement priorities: It is important not to inflate your performance assessment system with irrelevant indicators. If externally imposed targets are important to your organization, then they should be part of your strategy. However, if they are just things you need to report on, then separate them out from the internal measurement system and tell everybody the reason why they are being collected. If we just dump any indicators into our local measurement system, we dilute it, and with it, we loose relevance and focus.
- Do identify KPQs before you start collecting any indicators: In order to measure something, it is important, first, to know what exactly you are talking about here and why you care about measuring it.[39] If we haven't got a real need for information, then we shouldn't need any indicators. KPQs ensure that your information needs are clarified and clearly articulated.
- Don't measure just for the sake of measurement: We have become experts in collecting and storing data. The problem is that many organizations focus all their attention on the actual activities related to measurement and forget to do something with the data they are collecting. This can produce an institutionalized and bureaucratized system of measurement that people will detest.
- Don't just rely on numeric data: Quantitative approaches alone cannot answer all our questions about performance.[40] Dee Hook, founder of the Visa network, said, 'In years ahead, we must get beyond numbers and the language of mathematics to understand, evaluate and account for such intangibles as learning, intellectual capital, community, beliefs and principles, or the stories we tell of our tribe's values and prosperity will be increasingly false.'[41] Measurement isn't primarily about calculation, data collection and analysis; it's not just a static, technical process, but one of ongoing interaction.
- Do create honesty and trust: Trust is an important ingredient of successful performance assessment. Trust can be achieved only through honesty when it comes to indicators and performance assessment. Measurement doesn't have to be perfect to be trusted. However, there must be honesty and any limitations of measurement must be openly acknowledged.

- Do create an environment in which people feel in control of measurement: Ownership of measurement is an important success factor as it creates understanding and trust. People will engage in the process if they feel that they are in charge of measurement, as opposed to the feeling that measurement is done to them. For this it is important to involve people across the organization in the design of performance indicators. The onus is often on senior and middle managers to let go and allow their subordinates to get on with measuring the things they feel are important.

- Don't wait for perfect indicators: It is important to acknowledge that imperfect measures are fine – as long as we acknowledge the fact. There is no perfect measure. As long as indicators help us to get to the information we need and as long as they reduce uncertainty to help us make better decisions, this is enough.

- Do encourage people to experiment with new performance indicators: We have to create an environment in which people aren't afraid of trying something new and innovative when it comes to measurement. People should have the freedom to question existing performance indicators and experiment with new ones. In fact, the process of assessing is a learning process – which should be used to improve the indicators used to assess performance.

- Do use the performance indicators and performance assessments to interact with people: When it comes to performance assessment, interaction between organizational members is critical to success. This allows an honest exchange of performance and creates trust across the organization. For example, managers use face-to-face discussions to engage people in a dialogue about performance. They clarify what matters and motivate information gathering outside of any routine channels.[42]

- Do use performance indicators to learn: Only if performance assessment and performance indicators help us improve future performance, they are of any value. Indicators should help you gain new insights that lead to better decision making and learning. Learning is the basis for any improvement and therefore should be the main focus of any performance assessment activities.

- Do manage the tensions between the different measurement usages: Even if you have created a measurement system that is focused on learning, there is a danger that after a while people shift to controlling and reporting functions. This requires continuous management attention and efforts to ensure indicators are not used in a command-and-control fashion, as this destroys trust. Also, we get too easily dragged down into the process of collecting and reporting indicators and need to make sure we are keeping our head above the water.

- Do apply common sense: Most importantly, we have to apply common sense when it comes to performance assessment. Unfortunately, common sense is not always common practice, and organizational routines and political pressures can get in the way. Performance assessment is best done in the way we as human beings routinely use it to interact and make sense out of the world around us in our daily life.

SUMMARY

- In this chapter, I have discussed the important role of performance measurement. Measurement is an essential part of human interaction, the foundation of science and progress and something we as human beings do routinely in our daily routines.
- Performance measurement plays an essential role in organizations and is too important for senior managers to ignore and leave to technical specialists.
- There are different reasons why we use measurements in our organizations. We can use measures to *control people's behavior*, for *external reporting and compliance* or for *learning and empowerment.*
  - ○ Controlling behavior: Measures are used in a top-down command-and-control fashion to guide and control people's behavior and actions.
  - ○ External reporting and compliance: Measures are used to inform external stakeholders and to comply with external reporting regulations and information requests.
  - ○ Learning and empowerment: Measures are used to empower employees and to equip them with the information they need to learn and make decisions that lead to improvements.
- When it comes to social organizations, we reach many limitations of measurement and *cannot design perfect* and all-encompassing *measurement systems*. This leaves *blind spots* in our measurement systems.
- These limitations have *severe implications for the way measures are used.* The command-and-control model breaks down completely and, if applied, opens the doors for dysfunctional behavior, data manipulations and cheating.
- When it comes to collecting measures for external reporting and compliance, we have to be careful that externally imposed metrics do not *unnecessarily inflate* the organizational measurement system. Indicators for external reporting that are not clearly linked to the strategy of the organization have to be kept separate.
- Learning and empowerment should be the main focus of any performance measurement activities. Here indicators are used to gain insights that help us make better-informed decisions and improve performance.
- I suggest using the word *performance indicator* rather than performance measure because our indicators only indicate performance but don't measure it. A performance indicator identifies to what degree a variable is present.
- I also suggest using the word *performance assessment* instead of performance measurement. Assessment is broader and allows taking into account information sources other than just quantitative data. The process of performance assessment uses indicators to reduce uncertainty and to compare a given situation or status relative to known objectives or goals.

- Performance assessment should enable and empower everybody within an organization to evaluate performance and gain insights that lead to better decision making and performance improvement.
- Finally, I introduced a number of important dos and don'ts of performance assessment in government, public sector and not-for-profit organizations.

REFERENCES AND ENDNOTES

1. According to Klein, H. A. (1974). *The Science of Measurement: A Historical Survey.* Dover Publications, New York

2. For a more exhaustive list, see for example: Whitelaw, I. (2007). *A Measure of all Things: The Story of Measurement through the Ages.* David & Charles, Cincinnati, OH; Robinson, A. (2007). *The Story of Measurement.* Thames & Hudson, London

3. See for example: Robinson, A. (2007). *The Story of Measurement.* Thames & Hudson, London

4. This sentence is based on: Spitzer, D. R. (2007). *Transforming Performance Measurement: Rethinking the Way We Measure and Drive Organizational Success.* American Management Association, New York, p. 22

5. See also the idea of cybernetic control as discussed in Hofstede, G. (1978). The Poverty of Management Control. *Academy of Management Review,* 3(3), 450–461; the concept of diagnostic control as developed by Simons, R. (1995). *Levers of Control: How Managers Use Innovative Control Systems to Drive Strategic Renewal.* Harvard Business School Press, Boston; and the concept of motivational measurement as discussed by Austin, R. D. (1996). *Measuring and Managing Performance in Organizations.* Dorset House Publishing, New York, p. 193

6. See also the concept of homeostatic control as discussed in Hofstede, G. (1978). The Poverty of Management Control. *Academy of Management Review,* 3(3), 450–461; the idea of interactive control as developed by Simons, R. (1995). *Levers of Control: How Managers Use Innovative Control Systems to Drive Strategic Renewal.* Harvard Business School Press, Boston; and the concept of informational measurement as discussed by Austin, R. D. (1996). *Measuring and Managing Performance in Organizations.* Dorset House Publishing, New York, p. 193

7. To learn about multiple intelligences, see for example: Gardner, H. (1993). *Multiple Intelligences: The Theory in Practice.* Basic, New York; Gardner, H. (2000). *Intelligence Reframed: Multiple Intelligences for the 21st Century.* Basic, New York

8. To find out about emotional intelligence, see for example: Goleman, D., Boyatzis, R. and McKee, A. (2001). Primal Leadership: The Hidden Driver of Great Performance. *Harvard Business Review,* 79(11), 42–51; Goleman, D. (1996). *Emotional Intelligence: Why It Can Matter More Than IQ.* Bloomsbury, London

9. From an interview with Daniel Yankelovich, quoted in Adam Smith [pseudonym of George J. W. Goodman] (1973). *Supermoney.* Michael Joseph, London, p. 286

10. de Bruijn, H. (2007). *Managing Performance in the Public Sector,* 2nd ed. Routledge, London.

11. Porter, T. M. (1995). *Trust in Numbers: The Pursuit of Objectivity in Science and Public Life.* Princeton University Press, Princeton, NJ

12. Boyle, D. Ibid (2001). *The Sum of Our Discontent: Why Numbers Make Us Irrational,* p. 38. Texere, New York

13. See for example: Gooday, G. J. N. (2004). The Morals of Measurement: Accuracy, Irony, and Trust in Late Victorian Electrical Practice. Cambridge University Press, Cambridge; Porter, T. M. (1995). Trust in Numbers: The Pursuit of Objectivity in Science and Public Life. Princeton University Press, Princeton, NJ

14. See: Austin, R. D. (1996). *Measuring and Managing Performance in Organizations*. Dorset House Publishing, New York, p. 13

15. Sugarman, C. (1990). *US Produce Standards Focus More on Appearance Than Quality*, The Pittsburgh Press. August 5, p. E1

16. See for example: Kerr, S. (1975). On the Folly of Rewarding A, While Hoping for B. *Academy of Management Journal*, 18(4), 769–783

17. See for example: Taylor, F. (1913). *The Principles of Scientific Management*. Reprinted in 2003 by Lightning Source UK Limited, Buckinghamshire, UK

18. Meyer, M. W. (2002). *Rethinking Performance Measurement – Beyond the Balanced Scorecard*, Cambridge University Press, Cambridge. p. xxi

19. For more information and case study material, see also: Marr, B. (2005). *Strategic Performance Management: Lessons from Call Centres*. Cranfield School of Management, Cranfield, UK

20. Another classic example is referral interview in a government agency; here the number of interviews is measured, whereas the quality of referrals is not. See: Blau, P. M. (1963). *The Dynamics of Bureaucracy: A Study of Interpersonal Relations in Two Government Agencies*. University of Chicago Press, Chicago

21. Ridgway, V. F. (1956). Dysfunctional Consequences of Performance Measurements. *Administrative Science Quarterly*, 1(2), 240–247

22. Meyer, M. W. (2002). *Rethinking Performance Measurement – Beyond the Balanced Scorecard*, Cambridge University Press, Cambridge. p. 8

23. Ehin, C. (2000). *Unleashing Intellectual Capital*, Butterworth Heinemann, Boston. p. 138

24. Ehin, C. (2000). *Unleashing Intellectual Capital*, . Butterworth Heinemann, Boston. For quote, see page 179

25. See for example: Torres, R. T., Preskill, H. S. and Piontek, M. E. (1996). *Evaluation Strategies for Communicating and Reporting: Enhancing Learning in Organizations*. Sage, Thousand Oaks, CA, p. 2

26. To view the performance statistics for New York City, visit: http://www.nyc.gov/html/ops/cpr/html/cpr_home/cpr_home.shtml

27. See: http://vaperforms.virginia.gov/

28. Source: http://www.nyc.gov/html/ops/cpr/html/about/about.shtml

29. See also: Spitzer, D. R. (2007). Transforming Performance Measurement: Rethinking the Way We Measure and Drive Organizational Success. American Management Association, New York, p. 30

30. See for example: Power, M. (1994). *The Audit Explosion*. Demos, London

31. See for example: Power, M. (1999). *The Audit Society: Rituals of Verification*. Oxford University Press, Oxford, p. 1

32. See for example: Austin, R. D. (1996). *Measuring and Managing Performance in Organizations*. Dorset House Publishing, New York, p. 123

33. See for example the idea of New Public Management: Hood, C. (1991). A Public Management for all Seasons? *Public Administration*, 69(1), 3–19; Osborne, D. and Gaebler, T. (1992). *Reinventing Government*. Addison Wesley, Reading, MA

34. See for example: Spitzer, D. R. (2007). Transforming Performance Measurement: Rethinking the Way We Measure and Drive Organizational Success. American Management Association, New York, p. 3

35. Kelley, D. L. (1999). *Measurement Made Accessible: A Research Approach Using Qualitative, Quantitative, & Quality Improvement Methods*, Thousand Oaks CA. p. 2. Sage

36. See for example: Hubbard, D. (2007). *How to Measure Anything: Finding the Value of Intangibles in Business*. Wiley, Hoboken, NJ, p. 21; who also uses the idea of reducing uncertainty

37. See for example: Bevan, G. and Hood, C. (2006). What's Measured Is What Matters: Targets and Gaming in the English Public Health Care System. *Public Administration*, 84(3), 517–538

38. de Bruijn, H. (2007). *Managing Performance in the Public Sector*, 2nd ed. Routledge, London. pp. 44 and 62

39. See for example: Hubbard, D. (2007). How to Measure Anything: Finding the Value of Intangibles in Business. Wiley, Hoboken, NJ, p. 25

40. See: Smith, P. C. (2006). Quantitative Approaches Towards Assessing Organizational Performance. In: *Public Service Performance: Perspectives on Measurement and Management* (Boyne, G., Meier, K., O'Toole, L. and Walker, R. eds). Cambridge University Press, Cambridge, p. 90

41. Quoted in Boyle (2001), p. 29 (see note 12 above)

42. See for example: Simons, R. (2005). Levers of Organization Design: How Managers Use Accountability Systems for Greater Performance and Commitment. Harvard Business School Press, Boston; Simons, R. (1995). Levers of Control: How Managers Use Innovative Control Systems to Drive Strategic Renewal. Harvard Business School Press, Boston, p. 96

Chapter 7

Creating Key Performance Questions

The French philosopher Voltaire once advised to 'judge of a man by his questions rather than by his answers' and Albert Einstein maintained that all he ever did was asked simple questions. If that was good enough for Einstein to turn the study of physics on its head, then it should certainly be good enough for those of us who want to bring some fresh thinking to the subject of strategic performance management. Questions addressed in this chapter include:

- What are key performance questions (KPQs)?
- Why are KPQs important for performance management?
- How can we use KPQs to design better performance indicators (PIs)?
- How can KPQs help us to interpret performance information?
- How do we design KPQs in practice?
- What examples of KPQs are used by other government, public sector and not-for-profit organizations?

In the previous chapter, we discussed the role of measurement. Indicators are designed to provide us with answers. KPQs, on the other hand, are designed to identify the most important unanswered questions in relation to performance. Questions help us identify and articulate our information needs and trigger a search for answers.

Too often, do we focus on finding answers without asking the right questions. Nobel Prize winner Paul A. Samuelson makes a point when he quite rightly says, 'Good questions outrank easy answers.' In most government, public sector and not-for-profit organizations, we spend too much time and effort on finding answers and not enough time on asking the right questions. The concept of KPQs was developed to change this.

KPQs are a new and powerful innovation in the field of corporate performance management.[1] A KPQ is a management question that captures exactly what it is that people want to know when it comes to organizational performance and each of the strategic elements and objectives on the value creation map (VCM). The rationale for KPQs is that they provide guidance for collecting relevant and meaningful PIs and focus our attention on what actually needs to be discussed when we review performance. Far too often do we jump straight to designing indicators before we are clear about what it is that we want to know.

By first designing KPQs, we are able to ask ourselves: 'What do we really need to know?' 'What information do we require?' and 'What are therefore the best PIs we need to collect to help us answer our key performance questions?'

Starting with KPQs ensures that, by default, all subsequently designed PIs are relevant and address real information needs. In addition, KPQs put performance data into context and therefore facilitate communication, guide discussion and direct decision making.

QUESTIONS HELP US TO LEARN

It is important to remember that the main reason for strategic performance management is to improve future performance. Performance improvement is based on learning. Deep and significant learning occurs only as a result of reflection, and reflection is not possible without a question. KPQs are therefore essential components of good performance management. KPQs allow us to put performance information and data into context and turn it into knowledge. Data and information contained in PIs are not useful on their own and cannot be turned into knowledge or wisdom unless we have questions we want to answer. Once we have got a question, we can then use data to turn it into knowledge and learning (see Fig. 7.1). Without questions there can be no learning, and without learning there can be no improvement.

Many government, public sector and not-for-profit organizations now apply KPQs. An example of how powerful KPQs can be in strategic performance management comes from Google – one of today's most successful and most admired companies on the planet. Google applies the principles of KPQs, and CEO Eric Schmidt says[2]:

> We run the company by questions, not by answers. So in the strategy process we've so far formulated 30 questions that we have to answer […] You ask it as a question, rather than a pithy answer, and that stimulates conversation. Out of the conversation comes innovation. Innovation is not something that I just wake up one day and say 'I want to innovate.' I think you get a better innovative culture if you ask it as a question.

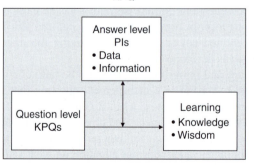

FIGURE 7.1   KPQs, PIs and learning.

A MORE SYSTEMATIC APPROACH TO DATA COLLECTION

Any student of science learns that it is important to know what you are look-ing for before you start collecting any data. The basics of the scientific method are that we first need to define a question we want to explore, we then collect more information and form a hypothesis. After that you decide on the most appropriate research method and collect the data, which we then analyze and interpret to draw conclusions about the question you set out to answer.

If we start collecting data without knowing what answers we are looking for, then we often end up collecting wrong or unnecessary data and are left with few or no real insights about the important questions we need answers to. This is a massive and universal problem that I have observed in every organi-zation I worked with. A typical example is when an organization wants to understand whether its employees, customers or partners are happy, the man-agement team often goes straight to the indicators and forms a project team that looks for the best ways of measuring this. However, it ends up looking for generic and 'proven' ways of collecting such information and often adopts existing surveys.

Of course, it makes sense to build on what experts have developed over the years. However, in our desire to find measures and get our hands on the data, we often fail to clarify what it is we really want to know. For example, once we have decided that the relationships with our partners are important and we ought to measure it, we need to pause and clarify what it is we want to understand. Here is where KPQs come in – defining the question or questions we want to have an answer to forces us to be more specific and spell out what it is we want to know. Once we have the question, we then have to ask ourselves: What is the information we require that will help us answer this question and what is the best way of collecting this information?

Let me give you an example. A major organization approached me to audit its performance management approach, which is something the Advanced Performance Institute does regularly with its clients. So we spent a few days with the client to really understand its performance management and perform-ance measurement approach. As part of its strategy, this organization had moved to a partnership model, and for that reason it was critical to success-fully manage the various partnerships it had to deliver joint outcomes. When this organization moved to its partnership-based delivery model a few years back, the organization wanted to find ways of measuring and assessing its part-nerships. In a quest to find measures, it came across a company which special-ized in partnership evaluations and which had designed a generic questionnaire to measure partnerships. The project team in the organization was pleased about this. It signed up to this survey and outsourced its data collection to this outside company which then started to collect the partnership data twice a year. Again, managers in the organization were pleased with the service they received from this external company. It provided them with detailed reports containing graphs, tables and trend analyses on about 50 different questions contained in the survey.

While the managers in the organization were happy with how things were going with their partnership assessments, the partners were telling a different story. When I spoke to some of the key partners, it became apparent very quickly that they were not happy with the way the partnerships were assessed and how the data was collected. One manager of a partner organization told me:

> …of course we want to ensure that we create a good relationship. However, I am getting really annoyed with them. Twice a year they send me a six-page long survey which I need to complete. To collect all the information takes me about 3 days and is a lot of work. The problem is that I can't see why they need all of this data, a lot of the questions seem completely irrelevant to our partnership.

After speaking to their partners, I went back to the organization and asked the managers a few more questions about how they were using the information they were collecting in their surveys and how this was facilitating their decision making. It very quickly became clear that all of the data they were collecting was 'interesting to know', but that was it. Not one single decision had been taken based on the survey data over the past 3 years. So, in conclusion, they were creating a lot of unnecessary work for themselves and most importantly for their partners, which started to undermine the very relationship they were trying to assess and improve. This example is not a one-off – I frequently see similar problems in organizations all over the world. Just think of your staff satisfaction surveys! So how can KPQs help?

HOW QUESTIONS CAN GUIDE US TO THE RIGHT INDICATORS

Let's go back to the example of partnership assessments. When the managers of the organization realized what they were doing and how this process was creating all this unnecessary work for no real value, they went back to the drawing board and identified the question(s) they really wanted to have an answer to. The KPQ they came up with is "How well are our partnerships progressing?" Once they had a KPQ, they asked themselves what data they would need to answer this question and what would be the best method to collect the data? They needed data that would assess the relationships but they didn't want to use the same survey again as this was collecting too much unnecessary data.

After some deliberation, they agreed that the best way forward would be to ask their relationship managers or account managers for an assessment. The members of the project team realized that along with the account managers they had people in place who would be able to make a comprehensive assessment without the need for a lengthy survey. They designed a system that automatically e-mailed a very simple form to the account managers with just two questions: 'How would you assess the relationship with partner organization X?' and 'How well is the partnership with organization X progressing?' Next

How would you assess the relationship with company X?	Problematic	Indifferent	Positive
	◯	◯	◯
Written comment:			
How well is the partnership with company X progressing?	Worse than before	Same as before	Better than before
	◯	◯	◯
Written comment:			

FIGURE 7.2   Assessing partnership performance.

to the question the form included a scale. Initially, this was a 10-point scale from very bad (0) to very good (10). This was later refined into a 3-point scale. In addition to the scale, the form also included a field for a written comment (see Fig. 7.2). Account managers were now asked to assess the partnerships by ticking a box on a scale and by providing a short written comment on why they picked that particular assessment.

The project team members realized that by asking only the account managers they might get a biased view on the situation since they were only collecting internal data. So they decided to also e-mail the form to their partner companies. They also had people in charge of managing the relationship who were perfect candidates to provide the external view. Preferring not to ask for any written assessment, they used a form for the partner organization that only included the two scaled questions. Once account managers and the partner company had completed the short survey, the results were stored in a database and were automatically compared. In over 95% of the cases, the internal and external assessments were identical. Where major differences in opinion occurred – let's say the internal person thought everything was fine and the external partner felt there were problems – then the database triggered another e-mail to the internal account manager prompting him or her to pick up the phone and discuss any potential issues with the partner organization. After a while, the project team members also realized that they were not collecting such data frequently enough. Getting such information only every 6 months meant that potential problems could develop undetected for a long time. They decided that monthly data was required in order to be able to react to potential issues early enough before they became big problems. This organization now has a very simple monthly data collection system in place, which allows it to get all the information needed to answer its KPQs. All of this in turn helps it to manage the partner network to deliver the joint outcomes.

Carl Sagan writes, 'We make our world significant by the courage of our questions and by the depth of our answers.'[3] Having the right questions in place allows us to collect the right management information so that we can produce deep and meaningful answers. In Chapter 8, I discuss in much more detail how to design key performance indicators (KPIs) based on your KPQs.

QUESTIONS HELP US TO INTERPRET

While KPQs help us to create better and more meaningful PIs, they do much more. KPQs are able to put management information and performance data into context. When we communicate PIs and management information in reports or tables, we often fail to communicate why this data is important and what it helps us to understand. However, if we ensure we communicate KPQs with the management information and performance data, it allows a recipient to understand why any data is relevant and what questions it helps to answer. It also allows us to make a judgment about how useful and comprehensive the performance information is in helping us to answer our KPQs.

Far too often do we circulate information without providing the relevant context, and far too often do we expect people to collect data and information without a question that puts it into context and explains why we need the performance data. I advise my clients to always circulate KPQs with any data they are reporting or collecting in order to provide reference points and context.

KPQs are also powerful tools to evaluate your existing set of PIs. If you ever feel that you are collecting a lot of seemingly meaningless information in your organization, then just ask the question: 'What is the KPQ we are trying to answer with this data?' If, on the one hand, you are able to frame a KPQ for your existing PIs then this allows you to put this existing data into context and helps to clarify to users why you need to collect it. If, on the other hand, you are unable to identify a KPQ that any existing PIs is helping to answer, then you have just discovered seemingly unnecessary data that you are collecting, which needs to be challenged. KPQs can therefore be a great tool to review and improve any existing set of PIs.

THE POWER OF QUESTIONS

Questions have a physical, mental and emotional impact on humans. Once someone has asked us a question, it triggers a search mechanism in our brains. This is the start of a thinking process that allows us to reflect on possible answers, which constitutes the beginning of learning. KPQs indicate to everyone what is of most concern to the organization and the people in it. In fact, when managers and executives ask questions, they send the recipients of their questions on a mental journey or quest for answers. Donald Peterson, former CEO of Ford Motor Company, once said that 'asking more of the right questions reduced the need to have all the answers.'

In our society, we are expected to be decisive and know the answers to all the questions. We often feel defensive when people ask us questions or we hesitate to ask good questions fearing that might reveal that we don't know something. Many of us seem to have lost the ability to ask questions. This is strange as it is the very ability to ask questions that allows us to learn. Just think of the ease and curiosity with which young children ask questions. Human progress came about when man started to asked questions. Stagnation results not from a lack of answers but from the absence of meaningful questions.

Michael Marquardt, professor at George Washington University, makes the point that[4] 'In organizations that discourage questions, information is usually hoarded, people keep their heads down and stick to their knitting, and few people are willing to take any risks', he continues to stress that, 'These organizations usually suffer from low staff morale, poor teamwork, and poor leadership. They become fossilized, even moribund.' Management professor Sydney Finkelstein agrees and argues that organizations that are not able to ask the right questions constitute 'zombie companies': 'a walking corpse that just doesn't yet know that it's dead – because this company has created an insulated culture that systematically excludes any information that could contradict its reigning picture or reality.'[5]

I believe that KPQs are the critical link between your strategic objectives, PIs and performance improvement. Without the right questions it is hard to identify relevant and meaningful indicators, and without questions it is impossible to learn and improve. Knowledge and wisdom come from asking the right questions. When we ask questions instead of focusing only on collecting answers, we invite people into the discussion. We engage the recipient in a dialogue and in a search for answers. This in turn leads to reflection, learning and improvement. KPQs can be extremely powerful management tools as they:

- enable us to design meaningful PIs;
- wake us up and focus our attention;
- help us to think clearly, logically and strategically;
- allow us to challenge the status quo;
- build a culture of engagement and accountability;
- unlock a conversation and trigger a dialogue; and
- lead to reflection, new insights, learning and performance improvement.

While we often ask ourselves questions (often unconsciously), we have to make this process much more explicit and better articulate the KPQs in organizations for everyone to see. In the following sections, I will discuss how to create KPQs in practice.

HOW TO CREATE GOOD KPQs

Asking good questions is difficult but rewarding.[6] Below, I have created a number of steps or principles that will enable you to create good KPQs for your organization. While having any question is better than having no questions, the real power is only unleashed when KPQs are designed properly. The following 10 steps will ensure that you create good KPQs:

1. Design between one and three KPQs for each strategic objective on your VCM.
2. Ensure KPQs are performance related.

3. Engage people in the creation of your KPQs.
4. Create short and clear KPQs.
5. KPQs should be open questions.
6. KPQs should focus on the present and future.
7. Refine and improve your KPQs as you use them.
8. Use your KPQs to design relevant and meaningful PIs.
9. Use KPQs to refine and challenge existing PIs.
10. Use KPQs to report, communicate and review performance.

Let me now discuss each of these 10 steps in a little more detail and provide more practical advice for creating good KPQs.

Design between One and Three KPQs for Each Strategic Objective on Your VCM

KPQs should be based on what matters in your organization, that is, your strategy. Once you have clarified your strategic objectives and mapped them into a VCM, you can start designing KPQs. My suggestion is to design between one and three KPQs for each strategic objective or strategic element on your map. Obviously, the fewer KPQs you have the better it is, because every KPQ will trigger KPIs and we don't want to create an unnecessary administrative burden in collecting data that is not really needed (see Fig. 7.3).

Similarly to KPIs, it is far too easy to come up with a whole host of questions and this is why some time needs to be taken to identify which questions you really need to have an answer to. You need to reflect on what is it you need to know in order to make better-informed decisions? In the same way that your strategic objectives should be unique, your KPQs should be unique and reflect the requirements and foci of your organization at this point in time. KPQs need to be regularly reviewed to ensure they still reflect the latest information needs.

There is often an iteration between setting your strategic objectives and creating your KPQs. If you are struggling to define good KPQs for your strategic objectives, then this can mean that your strategic objectives are not well defined and not clear. The discussion about KPQ can often lead to a revision and refinement of your strategic objectives on your VCM.

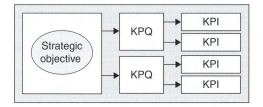

FIGURE 7.3 From strategic objectives to KPQs to KPIs.

Ensure KPQs are Performance Related

A KPQ needs to be about performance. The aim is to design questions you need to regularly revisit and answer in order to better manage your organization. Performance-related questions are those that allow you to understand how well you are implementing your strategic objectives and to what extent you are meeting your targets and objectives.

KPQs are not about strategic choice or strategy clarification! This is a trap many organizations fall into when they are first trying to apply KPQs. Instead of designing performance-related questions, they end up asking strategic choice questions such as 'should we deliver service X or service Y?' or strategy clarification questions such as 'When do we know our customers are happy?' (see also Table 7.1). Strategy-related questions should all be answered during your strategy definition and strategy mapping phase (see Chapters 1–4). If you end up with lots of strategic-choice questions or clarification questions about strategy, then this is a strong indicator that your current strategy is not clear and understood. In this case it would be advisable to go back and revisit your strategy.

Engage People in the Creation of Your KPQs

KPQs should not be designed in the boardroom alone. Designing KPQs is a great opportunity to engage everyone in the organization as well as some external stakeholders. Try to involve people in the process and ask them what question they would see as most relevant. Once you have designed a list of KPQs, take this back to the subject matter experts or different parts within and outside the organization to collect feedback.

For example, when designing KPQs that relate to marketing, involve the marketing experts in the process. Create a project team from your marketing

TABLE 7.1 Performance-related Versus Strategic-choice Questions

Performance-related KPQs	Strategic-choice or strategy-clarification questions
How well are we doing this?	How should we do this?
How well are we delivering the service X to customers Z?	Should we provide service X or service Y? Should we target customer A or customer B?
To what extent are the customers in market segment X likely to recommend our service to others?	How do we know whether our customers are happy?
To what extent are we using our budgeted manpower in area X effectively?	Do we allocate our resources appropriately?

experts and ask them to identify possible KPQs. Once a final number of KPQs has been agreed on by the management, take the KPQs that relate to marketing back to the marketing department to discuss and refine. Remember that KPQs communicate to everyone what really matters in an organization; the more people understand and agree with these questions, the more likely it is that everybody will be pulling in the same direction.

Create Short and Clear KPQs

A good KPQ is relatively short, clear and unambiguous. A KPQ should contain only one question. We often produce a string of questions that makes it much harder to guide meaningful and focused data collection. The language should be clear and must not contain any jargon or abbreviations that external people might not understand. Likewise, try to stay away from management buzzwords and ensure that the question is easy to understand and uses language that people in your organization can comfortably understand and use.

KPQs Should Be Open Questions

Questions can be divided into two types: closed questions and open questions (see Table 7.2 for examples). These two types of questions are actually very different in character, usage and response. A closed question seeks a short and specific response that can be provided with either a single word or a short phrase. Questions that can be answered with either 'yes' or 'no' are generally closed questions. Closed questions are easy to answer and often seek simple facts: what, when, where? They are closed because the control of the conversation remains with the questioner. An open question, on the other hand, 'opens the door' to the respondent and seeks an open-ended response. Open questions invite the respondent to think and reflect and provide explanations, opinions or feelings. Open questions often start with words such as what, why, how or describe. With open-ended questions, the questioner hands over control to the respondent.

TABLE 7.2 Open Versus Closed Questions

Closed Questions	Open Questions
Did you go on holiday this year?	What did you do on holiday?
Is this important to you?	Why is this so important to you?
Are our customers satisfied?	How well are we meeting our customer demands?
Are you happy with your current supplier?	Describe to what extent are we improving our supplier relationship?

Closed questions such as 'have we met our budget?' can be answered with a simple 'yes' or no' answer, without any further discussion or expansion on the issue. However, if we ask an open question such as 'how well are we managing our budget?', the question triggers a wider search for answers and seeks more than a 'yes' or 'no' response. Open questions make us reflect, they engage our brains to a much greater extent, and they invite explanations and ignite discussion and dialogue. Whenever possible, KPQs should be phrased as open questions.

Phrasing KPQs as open questions also ensures that the PIs that will be used to help answer the KPQs are not just accepted at face value. Instead, they will become the basis of evidence for a wider discussion and dialogue.

KPQs Should Focus on the Present and Future

Questions should be phrased in a way that addresses the present or future: 'To what extent are we increasing our market share?', instead of questions that point into the past, 'Has our market share increased?' By focusing on the future, we open up a dialogue that allows us to 'do' something about the future. We then look at data in a different light and try to understand what the data and management information means for the future. This helps with the interpretation of the data and ensures we collect data that helps to inform our decision making and performance improvement.

Refine and Improve Your KPQs as You Use Them

Once KPQs have been created, it is worth waiting to see what answers come back – that is, how well the PIs are providing answers to the questions and how well the KPQs help people to make better-informed decisions. Once they are in use, it is possible to refine them to improve the focus even more. This is a natural process of learning and refinement, and organizations should expect some significant change in the first 12 months of using KPQs. Experience has shown that after about 12 months the changes are less frequent and the KPQs become much better.

Use Your KPQs to Design Relevant and Meaningful Performance Indicators

Once you have designed a set of good KPQs linked to your strategic objectives following the above guidelines, you can use them to guide the design of meaningful and relevant PIs. The details of how to design PIs are discussed in Chapter 8.

Use KPQs to Refine and Challenge Existing Performance Indicators

KPQs can be used to challenge and refine any existing PIs. Especially any indicators that are imposed onto your organization by external stakeholders or regulators can be assessed. Linking them to your KPQs can allow you to put them into context and justify their relevance. However, if existing indicators can't be linked to your KPQs, then they need to be kept separate from your performance measurement system and treated as 'reporting only indicators'.

Use KPQs to Report, Communicate and Review Performance

KPQs can also be used to improve the reporting, communication and review of performance information by always putting the KPQs with the performance data we are presenting. This way, the person who looks at the data understands the purpose of why this data is being collected and is able to put it into context. Furthermore, it allows us to reflect on the answers. KPQs are therefore a good way to structure your performance review meetings. I will discuss the reporting and performance reviews in more detail in Part III of this book.

SOME PRACTICAL EXAMPLES OF KPQs

Below I have listed a selection of KPQs developed by organizations I have worked with over the years. These only serve as illustrative examples as each organization needs to design its own unique KPQs in order to reflect its unique strategic objectives and to address its own information needs. Also, please note that most of them have been generalized to make them more digestible and to protect the anonymity of the organizations that developed them.

Examples of KPQs:

- To what degree are the customers who receive our service likely to recommend us to others?
- How well are we shifting toward an innovative climate in our organization?
- To what level are our employees engaged?
- How well are we using the information we possess?
- To what degree are we reducing the stress levels among our employees?
- How does the outside world view the role of our organization?
- To what extent do we trust each other?
- How effective is the communications strategy?
- To what extent are we involved in the leading-edge debates in the gambling world?
- How well are we building active partnerships with our key stakeholders?
- To what scope are we successfully promoting our services?
- How well is knowledge shared within and between the directorates?
- To what degree does our nonspecialist training support our operational capability now, and in the future?
- To what extent are we responding to the most exciting income opportunities offered by the market?

- How well are we sharing one set of values of Y?
- To what degree are we optimizing our inventory?
- How well are we reducing the waste in area Z?
- To what extent are we achieving our cost-reduction target in area X?
- How well are we promoting our services?
- To what level are we enhancing our international reputation?
- How well are we innovating our service offerings?
- To what extent are we keeping our most profitable customers?
- How well are we communicating in our organization?
- To what degree are we continuing to work in teams?
- How well are we building our new competencies of X?
- To what extent are we retaining the talent in our organization?
- How well are we fostering a culture of continuous improvement?
- To what degree are we continuing to attract the right people?
- How well are we managing our allocated financial resources?
- To what extent do people feel passionate about working for our organization?
- How well are we helping to develop a coordinated network to perform service X?
- To what extent are our employees motivated?

The above list of real KPQs should give you a good feel for what KPQs might look like in practice. Next, I outline why I believe KPQs are as important, if not more important, than KPIs.

WHY KPQs CAN BE MORE IMPORTANT THAN KPIs

The US humorist and cartoonist James Thurber said that 'It is better to know some of the questions than all of the answers.' This couldn't be truer for KPQs. In my opinion, KPQs are actually more important than KPIs and constitute one of the most important innovations in the field of strategic performance management to date. The more I experience the benefits in organizations that use them, the more I am convinced that they are key components of successful strategic performance management initiatives.

When I first introduced KPQs, they were designed as tools to help us design better PIs. However, organizations have taken the application much further than that. Different organizations now produce scorecards exclusively based on their KPQs. In those cases, business analysts or managers discuss the KPQs in their performance review meeting, look at the evidence available to them in the form of PIs, and then decide on traffic lights or color codes for the KPQ. This allows managers to move beyond the often meaningless presentation and review of numbers and instead focus their thinking and dialogues on what really matters.

SUMMARY

- In this chapter I have introduced KPQs as a new and powerful innovation in the field of corporate performance management.
- Questions are the foundation for learning and improvement. Performance improvement is based on learning. Deep and significant learning occurs only as a result of reflection, and reflection is not possible without questions.
- KPQs allow us to identify what it is we want to know and therefore represent a critical link between our strategic objectives and the PIs we need to collect.
- KPQs articulate our information needs that in turn help us to identify appropriate and meaningful PIs.
- I have suggested to design between one and three KPQs for each strategic objective on your VCM.
- It was recommended to engage people in the creation of KPQs and to design more open questions which are performance-related, short, clear and focused on the present and the future. KPQs can be refined through their application and usage. Practical examples of KPQs were provided as illustrative examples.
- Besides using KPQs to design PIs, they can also be used to refine and challenge the existing PIs.
- Finally, KPQs can also be used to improve the reporting, communication and review of performance information. KPQs allow us to put PIs into context and guide our thinking process and reflection process. It therefore helps to focus our attention and dialogues to what matters the most.

REFERENCES AND ENDNOTES

1. The concept of Key Performance Questions was developed by Bernard Marr and the terms 'key performance question' and 'KPQ' are trademarks of the Advanced Performance Institute.
2. Eric Schmidt in an interview with Jeremy Caplan for TIME, October 02, 2006.
3. Carl Sagan (1980). Cosmos, Time Warner Books, London (page: 193).
4. Marquardt, M. (2005). *Leading with Questions: How Leaders Find the Right Solutions by Knowing What to Ask*. Jossey-Bass, San Francisco.
5. Finkelstein, S. (2004). Zombie businesses: How to learn from their mistakes. *Leader to Leader*, 32, 25–31 (p. 25).
6. Browne, N. and Keeley, S. (2007). *Asking the Right Questions: A Guide to Critical Thinking*. Pearson Prentice Hall, New Jersey.

Chapter 8

Designing Performance Indicators

Indicators are essential components of human existence; they represent the foundations of trade, science and progress. In our organizations, performance indicators are vital tools of management. Performance indicators provide us with the information and evidence that help us gain new insights, enable us to learn, assist us in our decision making and allow us to act on it to improve future performance. However, in many government, public sector and not-for-profit organizations, we make the fatal mistake to believe that every indicator is useful. The problem starts when we use performance indicators to just measure what is easily countable and when we end up becoming obsessed with measuring everything that walks and moves. This then causes all the negative consequences discussed in Chapter 6. In this chapter, I provide practical tools, frameworks and templates for designing relevant and meaningful performance indicators. The questions I will address in this chapter include:

- How do we design relevant and meaningful performance indicators?
- How many performance indicators do we need?
- How can a decision framework guide us to better indicators?
- What template can we use to develop performance indicators?
- How do we set the right targets?
- How do we ensure we collect the right data?
- What are examples of good and innovative performance indicators?

In our organizations, we need to start assessing what we value rather than value what is assessed. In an age of information overload, we need to distill down the data and information we collect to what is really relevant to the questions we have and to the decisions we need to make.

WHAT IS A PERFORMANCE INDICATOR?

In Chapter 6, I said that I prefer the word 'indicator' rather than measure, and I defined a performance indicator as something that allows us to identify to what degree a variable is present. Let me expand on this here.

- A performance indicator allows us to *collect evidence and information* that (1) helps us gain new insights and learning, (2) supports our decision making

175

and (3) leads to improved organizational performance. Indicators are the input that allows us to assess performance.

- It is important to reiterate that a performance indicator *does not necessarily mean counting* or quantifying. Evidence and information can also take the forms of written descriptions, observations, symbols, color codes, etc.
- A performance indicator has to be *relevant* to the information needs of the organization. It therefore (1) has to be linked to the strategic priorities of the organization, (2) has to relate to the important and unanswered questions in an organization and (3) has to be connected to the important decisions that have to be made in an organization.
- A performance indicator has to be *meaningful*. It therefore has to provide the right evidence and information to actually help answer our key performance questions (KPQs) and enable us to make decisions. We therefore have to collect the right information, from the right source, at the right frequency, and we have to provide the evidence and information to the right people, in the right format, at the right time.

If we fail to make our performance information relevant and meaningful, then it just becomes performance data that represents unnecessary 'noise' and which distracts us from learning, decision making and performance improvement.

COLLECTING INDICATORS AND ASSESSING PERFORMANCE IS EASIER THAN YOU THINK

We often think that assessing performance or collecting indicators means meeting some nearly unachievable criteria.[1] Also, we often feel the process of collecting and analyzing performance information should be done by specialists who have a mathematical or statistical background and who are trained in data collection and analysis. But this is not the case. Designing and collecting performance indicators and assessing performance are not difficult activities and everybody in the organization can do them.

In his book on performance measurement, Douglas Hubbard correctly provides four useful assumptions when it comes to performance indicators and performance assessment:[2]

1. Your problem is not as unique as you think.
2. You have more data than you think.
3. You need less data than you think.
4. There is a useful indicator that is much simpler than you think.

We also don't need 'perfect' indicators. We have seen in Chapter 6 that perfect indicators don't really exist. Performance indicators, therefore, don't have to provide us with complete answers to our KPQs and eliminate uncertainty totally; a mere reduction in uncertainty and the extraction of new insights is good enough.

INDIRECT AND PROXY INDICATORS ARE OKAY

We must remember that measures were made for man and not man for measures.[3] The performance indicators we use in our organizations are all indirect and proxy indicators – and that is okay. We often wrongly feel that some fields of science have perfect indicators. The fact is that they just have indicators that have been fine tuned over a longer period and therefore are proxy indicators that more reliably reflect reality. It is natural that areas of measurement evolve and improve over time and more generally accepted methods would surface. While Louis Pasteur, the famous French chemist and microbiologist, said that 'a science is as mature as its measurement tools', Bertrand Russell, British mathematician and philosopher, also maintained that 'although this may seem a paradox, all science is based on the idea of approximation. If a man tells you he knows a thing exactly, then you can be safe in inferring that you are speaking to an inexact man'.

For example, temperature was considered very qualitative and immeasurable until Daniel Fahrenheit developed the mercury thermometer to measure it. However, it still uses a proxy of the expanding mercury to indicate temperature. Today, we all accept this as a valid form of measurement for temperature. Time represents another good example. Of the many things that we measure, time is the one we are probably most aware of in our daily lives. We all have access to clocks and understand the units of time (seconds, minutes, hours, days, weeks, months, years).[4] The first bit of evidence that man was measuring time relates back to 37 000 years to the Paleolithic Period. By about 3000 bc, the Sumerians had refined calendars based on the observations of lunar cycles, seasons, and so on. While most civilizations have now adopted a solar calendar of 365 days, the actual time taken for the earth to orbit the sun or the time between the phases of the moon is not perfectly divisible by a whole number of days.[5] This is why we have to adjust our calendars by adding another day every 4 years to synchronize our measures of time with the actual movement of the earth around the sun. This leap year correction provides better alignment but is still not 'perfect'. Today, scientists have produced ever more precise definitions of our smaller units of time. In 1960, for example, the General Conference on Weights and Measures ratified a definition of the second as 1/31 556 925.974 7 of the length of the tropical year of 1900. It is only more recently that we have linked the definition of a second to a natural and fixed phenomenon. Scientists had discovered that the transition of an atom or molecule between two energy levels was unchanging and could be reproduced accurately anywhere. Therefore, we now have a very precise definition of a second based on 'the duration of 9 192 631 770 periods of radiation corresponding to the transition between the two hyperfine levels of the ground state of the cesium-133 atom'. We can now measure a second very precisely; however, the whole system of dividing the movement of the earth into years, months, weeks, days, hours, minutes and seconds is still only our best approximation.

The point I am trying to make here is that even measurements of something we believe to be 100% precise, such as time or temperature, are based on man-made proxies. The difference is that they started off as very basic methods but evolved into more sophisticated instruments over time.

Also, the instruments have become more precise based on our needs. Five thousand years ago, it wasn't necessary for farmers to have clearly defined seconds, minutes and hours. As the society was based on agriculture, a rough understanding of the seasons and the months was sufficient. However, in today's digital world, we require much more precise synchronizations and definition of time. So a second point I am trying to make here is that we need appropriate accuracy based on our information needs and requirements.[6]

Douglas Hubbard maintains that some amount of error is unavoidable in our organizational performance assessments. However, as long as resultant evidence and information is an improvement on prior knowledge and helps us to reduce uncertainty, then it is valid. In fact, some amount of error is central to how experiments, surveys and other scientific measurements are performed. However, in the scientific world we are much better at admitting the measurement limitations and making the errors explicit. This is why results are reported in ranges, for example the average number of satisfied customers has increased between 12% and 18% (95% confidence interval) instead of single numbers that are often perceived as fact.[7]

If something really matters in our organization, it will be observable or detectable. Hubbard argues that even touchy-feely sounding things such as creativity, employee empowerment or strategy alignment have observable consequences if they matter at all.[8] And if something is observable or detectable, performance indicators can be designed to capture it.

PERFORMANCE INDICATOR DECISION FRAMEWORK

To facilitate the design of more relevant and meaningful performance indicators, I have designed a 10-step performance indicator decision framework (see Fig. 8.1). This framework takes you through a number of questions and decisions that will lead to better performance indicators. My recommendation is that you go through this template for every new performance indicator you intend to develop and that you try to apply this framework retrospectively to any existing indications you are using in your organization. Below, I will now go through this framework step-by-step and will discuss each of the steps in further detail.

Step 1: Which Strategic Element Do We Want to Measure?

As outlined in the previous chapters, we need to clearly articulate the strategic element we intend to assess. Any performance indicator has to be linked to

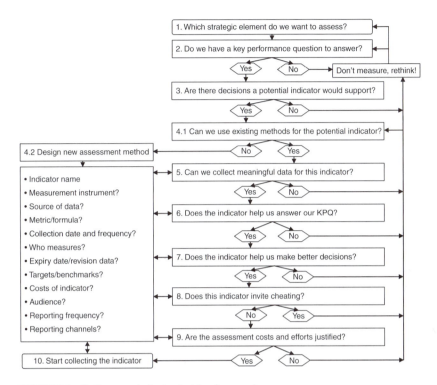

FIGURE 8.I Performance indicator decision framework.

our organizational priorities and strategic objectives. Our perceived inability to design indicators for the more complex or intangible elements of our strategy is not due to the fact that they can't be assessed. This perceived inability to measure is mostly due to the fact that we are unclear about what exactly we want to measure.

Step 2: Do We Have a Key Performance Question to Answer?

As outlined in Chapter 7, we need to clearly articulate our information needs by identifying a KPQ we want to have an answer to. By stating a question, we clarify why we want to measure anything. This therefore provides further context and narrows down the area of interest from a wider strategic objective to a narrower question. While we could probably come up with lots of indicators to measure our strategic objective, we set closer boundaries on our indicators by specifying our information needs. Where no KPQ can be identified, then there shouldn't be an indicator and users should try again to come up with a relevant question. If a KPQ can be identified, then the user can move on along the framework to the next question below.

Step 3: Are There Decisions a Potential Indicator Would Support?

While a KPQ narrows down the possible indicators that can be used, it still leaves many possible indicators to choose from. Another question can be used to narrow possible indicator down even further. This question is about clearly identifying any important decisions a potential indicator would help to make. By articulating the question and the possible decisions an indicator will help to address, it is possible to reduce the potential number of indicators from almost endless to a smaller and more focused set of possible indicators (see also Fig. 8.2). Where no decisions will be supported by a potential indicator, then it shouldn't be collected and users should try again to come up with a relevant decision or try again from the top of the framework. If possible decisions that this indicator will support can be identified, then the user can move on along the framework to the next question below.

Step 4.1: Can We Use Existing Methods for the Potential Indicator?

Before designing any new assessment and data collection instruments, it is important to check what has already been developed and used by others to

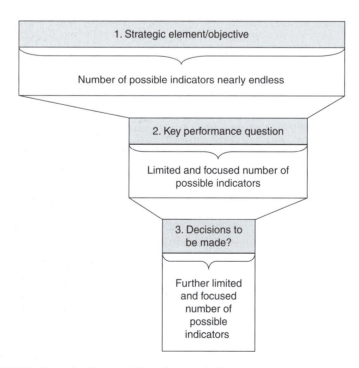

FIGURE 8.2   Narrowing down possible performance indicators.

avoid reinventing the wheel. In most cases, simple searches on the Internet will reveal a large number of possible methods that have been developed to measure performance. Some more in-depth searches using articles in industry-specific magazines or academic periodicals can usually reveal pros and cons for the different methods and some suggestions about when to apply which method. If no existing method can be found, which should be rare, then a new indicator has to be identified (Step 4.2, which will be discussed in more detail at the end of this section). If a seemingly suitable method for the potential indi-cator can be identified, then the user can move on along the framework to the next question below.

Step 5: Can We Collect Meaningful Data for this Indicator?

Even though an existing method is available and was initially chosen, it might not be possible to collect meaningful data. The source data might not be avail-able, data might not be available in the right format, and it might not be possi-ble to collect the data in the required frequency. If this is the case and we can't collect meaningful data for this indicator, then we have to go back to Step 4 of the framework and reevaluate excising and alternative data collection methods. If meaningful data can be collected for this potential indicator, then the user can move on along the framework to the next question below.

Step 6: Does the Indicator Help Us Answer Our Key Performance Question?

We might be able to collect meaningful data for our indicator, which can be collected in the right format and in the right frequency. However, it is impor-tant to add another sense check here and evaluate whether the indicator data is actually helping to answer the KPQs. It is sometimes possible to become so occupied with the process of designing indicators to forget the question it was designed to answer. If the indicator is not helping to answer the KPQ, then we have to again move up the framework to Step 4 and reevaluate excising and alternative data collection methods. Only if the indicator data is helping to answer the KPQ can the user move along the framework to the next question below.

Step 7: Does the Indicator Help Us Make Better Decisions?

This question is another sense check to ensure we can act on the data we are about to collect. Here, we evaluate whether the indicator data is actually help-ing us to make better-informed decisions. Too often do we end up with data that is interesting to know but which is not really enabling us to make better-informed decisions. Another problem is that the data that is collected can sometimes be ambiguous and therefore hard to interpret and act upon. If no decisions are made on the data, then no improvements will follow, and the indicators will be pretty worthless. If the indicator is not helping to inform your decision making, then we have to again move up the framework to Step 4 and reevaluate excising and alternative data collection methods. Only if the indicator data is helping us to make the decisions it was designed to support can the user move along the framework to the next question below.

Step 8: Will the Indicator Invite Cheating?

At this Step, it makes sense to reflect on the likelihood that the indicator and the targets associated with the indicator might invite people to cheat. It is useful to think about possible ways to deliver a good result and to hit the target without delivering good performance. The way some indicators are designed makes it easier to cheat than others. Take, for instance, the call center example discussed earlier. If we want to improve service delivery and measure call duration as a key measure that is tracked by the automated system, then this leaves considerable room for cheating because call center agents could just hang up – the system wouldn't know whether the call was handled satisfactorily or not. To avoid this, some call centers now ask customers to rate the call after the call has finished and the call center agent has hung up. This way the cheating can be avoided just by changing the way the data is collected. If the indicator makes it too easy to cheat, it would make sense to reflect on the indicator, redesign it or jump back to Step 4 to consider alternative data collection methods. Only if the indicator is fairly cheat-proof can the user move along the framework to the next question below. Please note again at this point that no indicator will ever be 100% cheat-proof (see also Chapter 6).

Step 9: Are the Assessment Costs and Efforts Justified?

Before starting to collect any indicator, it is important to consider whether the costs and efforts involved in collecting the performance indicator is justified. Collecting performance indicators can be very expensive. For example, the 'Best Value' measurement initiative introduced by the UK government was estimated to have added £29 million (almost US$ 60 million) a year to the cost of running the police force due to the extra efforts required for collecting and reporting against the different targets.[9] While some indicators are not very expensive to collect, other methods such as surveys and audits can be very expensive. The cost–benefit consideration ensures that money and efforts are only spent on the indicators that justify the costs. If the costs are too expensive and cannot be justified, then we have to again move up the framework to Step 4 and reevaluate excising and alternative data collection methods. Only if the incurred costs are justified can the user move along the framework to start collecting the data for the performance indicator.

Step 10: Start Collecting the Indicator

Once you have reached this 10th and final step of the framework for designing performance indicators, you can start collecting the indicator and rest assured that you have designed a relevant and meaningful performance indicator.

So far I have discussed the right-hand side of the 10-step framework for designing performance indicators (Fig. 8.1) and identified some of the key questions and decisions that will lead to better indicators. However, Step 4 splits into two phases: Step 4.1 for using an existing method and Step 4.2 on the left-hand side for designing a new assessment method. Step 4.2 includes the key elements of a performance indicator such as clearly defined measurement instrument, clearly defined source of the data, clearly defined measurement frequency, etc. In the following section, I will discuss the different elements of a performance indicator in more detail, which I have combined into a performance indicator design template. This performance indicator design template and the performance indicator decision framework are closely linked. As you can see in the framework diagram (Fig. 8.1), there is an interaction between Steps 5 to 9 and the indicator design elements (indicated by the double-sided arrows). This means that only by understanding the elements of a performance indicator can we answer the questions posed in Steps 5–9. Also, if we are not able to answer one of the questions in Steps 5–9 satisfactorily, we can sometimes tweak the design of the indicator (e.g. data source, assessment frequency, and so on) to improve the indicator.

PERFORMANCE INDICATOR DESIGN TEMPLATE[10]

To guide the indicator design, I have developed a performance indicator design template that can be completed in conjunction with the performance indicator decision framework described above (see Fig. 8.1). Completing this template ensures that organizations develop a sound and comprehensive understanding of each of their performance indicators. This is important because it ensures that all important elements on an indicator have been considered and that the data is consistently collected and interpreted. It eradicates the ambiguity, ambivalence and inconsistency that I see far too often with performance indicators. If indicators are to become the basis for decision making and learning, it is essential that everyone understands what these indicators mean, how reliable they are, where the data comes from, etc.

The template I present here can be used to develop completely new indicators or to develop a more comprehensive picture of existing performance indicators. The indicator design template clarifies why we need the indicator, provides information about how the data will be collected, identifies the targets set for the indicator and outlines who will see the data and in what format. Below, I will now explain each part of the performance indicator design template in more detail. For a summary of the template please see Fig. 8.3.

Performance Indicator Design Template	
The Basics – Why do we need this indicator?	
1. Indicator name	*Provide a clear indicator name.*
2. Strategic element being assessed	*State which strategic element of objective this indicator relates to.*
3. Owner of strategic element	*Identify the person(s) or function(s) responsible for the strategic element or objective.*
4. Key performance question	*Provide the key performance question this indicator is helping to answer.*
5. Decisions supported	*List the decision(s) this indicator is helping to support.*
How will the data be collected?	
6. Data collection method/ Instrument	*Describe the measurement instrument used and how the data is being collected.*
7. Source of data	*Identify where the data for this indicator comes from.*
8.Formula/Scale/Assessment	*Explain the scale or formula used to assess performance.*
9. Frequency	*Illustrate how often the indicator is measured.*
10. Who measures/ Reviews the data?	*Name the person who is collecting, updating and/or reviewing the data.*
11. Expiry/Revision date	*Identify until when this indicator will be collected or when itwill be revised.*
What are the targets?	
12. Targets/Performance thresholds	*Set the targets and/or benchmarks for this indicator and provide performance thresholds (e.g. for when traffic lights turn from green to amber to red, and so on).*
How good is the indicator?	
13. Cost Estimate	*Estimate of the costs incurred by introducing and maintaining this indicator.*
14. Confidence level	*Provide an evaluation of how well this indicator is measuring what it is supposed to measure (e.g. written comment/evaluation: e.g. good☺ fair ☺imperfect ☹).*
15. Possible dysfunctions	*Note down any possible ways this indicator could encourage cheating.*
Who will see the data? How will the data be presented?	
16. Audience/Access	*Identify the audience of the indicator data and outline who has access rights to it.*
17. Reporting frequency	*Illustrate when and how often the indicator is reported.*
18. Reporting channels	*Describe the channels used to report this indicator (reports, meetings, online, etc.).*
19. Reporting formats	*State how the performance data is presented (numerical, graphical, narrative formats)*
20. Notifications/Workflows	*Identify any notifications, e-mail alerts and workflows triggered by this indicator*

FIGURE 8.3   Performance indicator design template.

THE BASICS: WHY DO WE NEED THIS INDICATOR?

The first five elements of the performance indicator design template address the purpose of the indicator. The different elements have been discussed earlier and provide the necessary basics and the context for the indicator.

1. Indicator name: Any performance indicator needs a name that should clearly explain what the indicator is about.
2. Strategic element being assessed: The value creation map has identified the different strategic elements, objectives and priorities. Here, we identify which of those the measure relates to.

3. Ownership of strategic element: It identifies the person(s) or function(s) responsible for the management and delivery of the strategic element that is being assessed. This can be an individual employee or a team of people. The reason why it makes sense to clarify ownership here is to have someone who can be contacted in the future to discuss the performance or to fine-tune and improve the indicator.

4. Key performance question: Here, we identify the KPQ this indicator is helping to answer. This provides the context of why this indicator is being introduced and what the specific issue is that requires further information and evidence.

5. Decisions supported: Here, we list the decision(s) this indicator is helping to support. This provides further context and ensures we are clear about how we are planning to use the information and evidence provided by the indicator.

HOW WILL THE DATA BE COLLECTED?

In this part of indicator design template, we look at the more technical aspects of the data collection. Instead of just selecting any existing measurement method, it is important to consider the strengths, weaknesses, and appropriateness of different data collection methods.[11] Here, the designer of an indicator should include a brief description of the data collection method, specify the source of the data, how often the data is collected, what scale will be used to measure it, and who is in charge of collecting and updating the data.

6. Data collection method or measurement instrument: Here, we identify and describe the method by which the data is being collected. Examples of data collection methods or measurement instruments include surveys, questionnaires, interviews, focus groups, collection of archival data, etc. As discussed above, selecting the appropriate data collection method is important. I will discuss different measurement instruments in more detail in the section following this section on the design template.

7. Source of the data: Here, we identify where the data comes from. At this point, the designer of an indicator needs to think about the access to data and answer questions such as: Is the data readily available? Is it feasible to collect the data? Will the data collection method, for example interviews with senior managers, provide honest information? If not, maybe different data collection methods could be combined.

8. Formula/Scale/Assessment: Here, the designer of the indicators identifies how the data will be captured. Is it possible to create a formula? Is it an aggregated indicator or index that is composed of other indicators? Here, the designer also specifies if, for example, one of the following scales is used: nominal (numbering of categories, e.g. football players, or simple membership definitions, e.g. male or female); ordinal (determination

of greater or less, e.g. star rating for restaurants or movies); interval (determination of intervals, e.g. temperature in Fahrenheit or Celsius); or ratio (determination of equality and ratio in a continuum with a real zero, e.g. length, time, temperature in Kelvin); or whether the indicator is not expressed in any numerical form. The scale we pick will have implications on how we can use the data. For example, a nominal scale does not reveal any order or relative size, it just tells us whether something is one or the other. An ordinal scale allows us to understand that one is bigger or better than another, but doesn't tell us by how much. In addition to these classic scales, there is the Likert scale[12], which was designed to measure the extent to which respondents agree or disagree with statements. Each respondent is asked to rate a question or survey item on some response scale. The format of a typical five-level Likert scale is as follows:

1 = Strongly disagree
2 = Disagree
3 = Neither agree nor disagree
4 = Agree
5 = Strongly agree

There are different options to extend the classic 5-point scale and to change between odd-numbered and even-numbered scales. Odd-numbered scales have a middle value, which is a neutral response, labeled as 'neutral', 'undecided' or 'neither agree nor disagree'. Even scales on the other hand don't provide a neutral answer option and are so-called forced-choice response scales. Here, the respondents are forced to choose whether they lean more toward the 'agree' or 'disagree' end of the scale. Forced-choice questions are useful tools if there is reluctance among respondents to state their preferences.

9. Date and frequency of data collection: Here, the designer of an indicator thinks about when and how often the data for that indicator should be collected. Some indicators are collected continuously, others hourly, daily, monthly or even annually. It is important to think about what frequency provides sufficient data to answer the KPQs and helps to support the decisions outline in element five of this template. You wouldn't, for example, weigh yourself every hour of the day. To track body weight, for example, weekly weightings should be sufficient. Organizations might want to continuously track indicators for Web site usage or Web site downtime, while external indicators for brand ranking might only be available once or twice a year. One of the biggest pitfalls of performance assessments in organizations is that data is not collected frequently enough. For example, many organizations conduct employee survey once a year or even every 18 months. This is not very useful at all as the gaps between the assessments are too big and impacts of corrective actions cannot be tracked. Instead of surveying all

employees once a year, it is possible, for example, to survey a representative sample (let us say 10%) of employees every month so that individuals still only complete their survey once a year, but the organization receives monthly information that allows them to answer their KPQs and act on the data much quicker. It also makes sense to coordinate the dates when data is collected. Too much data collection is done on an ad hoc and uncoordinated manner so that bits of data are collected at different times. As a consequence of this, people in the organization might get too many and too fragmented requests for performance data. Furthermore, if different parts of an organization collect their data at different times and if different indicators are collected at different intervals, this makes it difficult or impossible to get valid snapshots of performance across different areas of the organization. Here, it is also sensible to understand the reporting requirements for this indicator (see element 17 of this template). If, for example, the data has to be reported at the end of the month, it then makes sense to collect the data in time to be able to analyze it, aggregate it, chase people who have not provided the data in time, solve any data issues or discrepancies and prepare the report.

10. Who measures and reviews the data: Here, we identify the person, function or external agency responsible for the data collection and data updates. The person responsible for measuring could be an internal person or function within your organization, or increasingly, it can be external agencies because many organizations outsource the collection of specific indicators. This is especially common for indicators such as customer satisfaction, reputation, brand awareness and employee satisfaction. As part of this element, we also clarify whether there are any review or sign-off cycles. It is common, for example, for one person to input the data and for another person to cross-check or sign-off the data before it is released.

11. Expiry or revision date: Indicators are sometimes introduced for a specific period of time only (e.g. for the duration of major projects or to keep an eye on restructuring efforts). The common practice is that a significant number of indicators are introduced once and collected forever because no one ever goes back and identifies the indicators that are not needed any more. Other obviously temporary indicators are introduced without giving them an expiration date; however, for those indicators a revision date should be set that allows the designers to review the template and check whether it is still valid. Even if indicators don't seem time specific, it makes sense to give all indicators revision date (e.g. in line with the annual planning cycle) to ensure they get reviewed and an assessment takes place to see whether they are still needed.

WHAT ARE THE TARGETS?

Every performance indicator needs a target or benchmark to put performance levels into context. In many government and not-for-profit organizations, the target-setting process is quite arbitrary and not enough thought is going into the setting of targets. Organizations often simply base it on previous performance figures and just suggest a target that looks 'a bit better' while others might simply calculate targets as mathematical 'steps', making fixed increases (e.g. 5% improvement), often with little thought as to how they will be achieved.[13] I have seen so many performance reviews where people have discussed the fact that customer satisfaction, for example, has dropped from 87% to 84%. However, nobody knows whether this is good or bad, whether this is in line with expectations, or how this compares to any targets or to the sector benchmarks. Target setting should not be viewed as an administrative process, but as an integrated and important part of designing meaningful performance indicators.

12. Targets and performance thresholds: The targets and performance thresholds identify the desired level of performance in a specified timeframe and put expected performance levels into context. A long history of research in goal-setting theory[14] and target-setting practice allows us to state that targets should be (1) specific and time bound, (2) stretching and aspirational but achievable and (3) based on good information. Many studies have shown that well-defined targets lead to a greater increase in performance, as opposed to generalized targets of 'do your best', which tend to lead to lower performance levels. Targets can be set as absolute targets (increase by 5), proportional or percentage targets (increase by 5%), relative to benchmarks (within the top three hospitals in our area or top quartile) or relative to costs or budgets (increase or decrease by 5% same level of budget). The following are a few tips for setting better targets:[15]
    - Use existing information and review trends and history.
    - Consider variations in performance, for example peaks, troughs and seasonal factors.
    - Take account of national targets, best practice benchmarks, and so on.
    - Take into account the cause-and-effect relationships, for example don't set top-level outcome targets before you have set appropriate targets for the enablers and inputs.
    - Take into account time lags (consider the value creation map and the time lags between the objectives).
    - Take into account any dependence on others such as partner bodies.

The following are some examples of good and poor targets:[16]

Good: 'We will reduce the number of missed household bin collections by 5% by next year.'
'We will cut the number of unfilled places in primary schools by 10% by December 31, 2010'
'We will increase the number of visits to local libraries by 20% before the end of 2015?'

Poor: 'We aim to have the best bus service in the region.'
'We will improve the way we handle complaints.'
'We will answer 75% of all letters within 5 days' (a poor target if the remaining 25% take 3 months to answer).

Many organizations use 'traffic lighting' to illustrate the levels of performance. Here, the designer of an indicator would therefore specify the thresholds for, for example, red/underperformance, amber/medium performance, green/ good performance and sometimes blue/overperformance. Here, it is also worth thinking about internal or external benchmarks; these can be derived from past performance, from other organizations or departments, or from forecasts.

HOW GOOD IS THE INDICATOR

By going through the performance indicator decision framework, we will have discussed the costs of measurement, the confidence level that this indicator is actually measuring what it is supposed to measure and any dysfunctions or cheating behavior this indicator might encourage. In this section of the indicator design template, the following things are captured.

13. Estimated costs: Another aspect that should be considered is the costs and efforts required to introduce and maintain a performance indicator. There is often an implicit assumption by many managers and measurement experts that creating and maintaining measurement systems does not incur significant costs.[17] However, on the contrary, measurement is expensive, especially if the indicators are supposed to be relevant and meaningful to aid decision making and learning.[18] Costs can include the administrative and/or outsourcing costs for collecting the data, as well as the efforts needed to analyze and report on the performance. It is important to ensure that the costs and efforts are justified.

14. Confidence level: Once the above aspects of an indicator have been addressed, it is time to think about the validity of the indicators. To what extent does the indicator enable us to answer the KPQ and support our decisions? For financial performance, the confidence level would normally be high because established tools are available to measure it. However, when we try to measure our intangibles, such as organizational culture, the confidence level would necessarily go down a peg or two. The assessment of the confidence level is subjective but forces anyone who designs an indicator to think about how well an indicator is actually 'measuring' what it was that it set out to 'measure'. Organizations have different preferences of how to express confidence levels; some use percentages (0–100%), others use grades (1–5; or low, medium, high), color codes (e.g. red, amber, green), or symbols (such as smiley faces). In addition, I suggested that a brief written comment is included to clarify the level of confidence and explain the limitations of an indicator.

15. Possible dysfunctions: Here, the designer of an indicator notes down any potential ways this indicator could encourage the wrong behavior or cheating. Reflecting on possible dysfunctions caused by indicators allows people to identify on possible better ways of collecting and assessing performance. In addition, it helps to raise the awareness of possible cheating behaviors that in turn enables everyone to monitor them much closer.

WHO WILL SEE THE DATA AND HOW WILL THE DATA BE PRESENTED?

In this final section of the indicator design template, the designer of an indicator identifies the way the performance indicator is reported. It identifies the audience, access restrictions, the reporting frequency, the reporting channels and reporting formats.

16. Audience and access: The designer of the indicator identifies who will receive the information on this performance indicator, as well as possible access restrictions. Indicators can have different audiences. It might therefore be a good idea to identify primary, secondary and tertiary audiences. The primary audience will be the people directly involved in the management and decision making related to the strategic element that is being assessed. The secondary audience could be other parts of the organization that would benefit from seeing the data. A possible tertiary audience could be external stakeholders. Also, audience groups have different functions and requirements. For example, some data will be provided to analysts who analyze the data further, while a management audience tend to need data to support their decision making.

17. Reporting frequency: Here, we identify how often this indicator is reported. If the indicator is to serve a decision-making purpose within the organization, then the indicator needs to provide timely information. The reporting frequency can be different from the measurement frequency. An indicator might be collected hourly, but then reported as part of a quarterly performance meeting. However, it is important to cross-check the reporting and measurement frequency to ensure they are aligned and that data is available.

18. Reporting channels: Here, the possible outlets or reports are identified, which are used to communicate the data. An indicator could, for example, be included in the monthly performance report to directors, could be presented in the bimonthly performance review meeting, could be included in the quarterly performance report to the board, could be included in the weekly performance reports to heads of service, could be reported on the organizational Intranet, or could be made available to external stakeholders through external reports or the Web site. It is again a good idea to cross-check the identified reporting channels with the reporting and measurement frequency to ensure they are aligned and that data is available in time.

19. Reporting formats: Here, we identify how the data is best presented. The indicator designer should clarify whether the indicator is reported as, for example, a number, a narrative, a table, a graph or a chart. The best results are usually achieved if performance is reported in a mix of numerical, graphical and narrative formats. Considerations should also include the presentation of a data series and past performance. A graph containing past performance might be very useful in order to analyze trends over time. This could also include targets and benchmarks. Increasingly too, organizations use traffic lights or speedometer dials to present performance data.

20. Notifications and workflows: Here, we identify proactive notifications and possible workflows. Workflows are predefined and automated business processes in which documents, information or tasks are passed from one person or group of persons to others. Notifications are predefined and automated messages and involve the proactive push of performance data, messages or alarm notifications to predefined individuals or groups. The definition of possible workflows and notifications is especially useful as a first step toward automation using performance management software applications. For example, e-mail notifications or workflows could be automatically triggered if an indicator requires an update, or to tell a specific audience that new data is available or whether an indicator has moved over or below a predefined threshold.

When designing any performance indicator, it is essential to constantly evaluate the validity and information value of the indicators. To what extent do the indicators enable us to assess the given strategic element? How well is this indicator helping us to answer the KPQ(s)? How well is the indicator supporting the outlined decisions? If the indicator is not providing us with the required information, we should not measure it at all.

DECIDING ON THE MEASUREMENT INSTRUMENT

An important step in designing indicators is to decide on the *measurement instrument* that will be used to collect the data. We often associate counting and using independent and archival sources with objectivity and reliability, and perception-based data with unreliability. This belief needs to change when it comes to performance assessment in our organizations. Also, while standard measures might help, the most benefits and insights come from the unique indicators that deliver insights on your unique value creation and your unique performance issues and questions. Soon, more measures will be qualitative and even quite subjective. For example, well-designed rating scales can often be worthwhile assessment tools for many aspects of performance that cannot yet be measured in a more objective way using standard methodologies.[19] We have to start taking qualitative and perception-based indicators seriously. Many studies have shown that perceptual assessments are as reliable, if not more reliable, as archival data.[20] Perception data can provide richer insights into the real level of performance and it allows us to actively involve people in the performance assessment. The way we can involve people is to ask them to, for example, rank competitors, evaluate the service delivery or organizational culture, assess the level of relationships with different suppliers, etc. These assessments can take the form of numerals, or grades; however, they can also take the form of traffic lights, symbols such as thumbs up or down, as well as written assessments. Written assessments are able to capture much more information and allow us to more naturally communicate assessment outcomes. If numerals are used to assess performance, it usually makes sense to supplement them with at least a comment field to provide some explanatory narrative assessment in addition to a number. Dean Spitzer, performance measurement thought leader for IBM, makes an important point in his excellent book on performance measurement when he argues that measurement is, at its roots, a social phenomenon, that is not a detached process of calculating abstract numbers.[21]

For example, instead of the ubiquitous and intrusive customer satisfaction survey, many service providers, such as hotels or banks, now use focus groups to identify what really matters to their target customers and then employ professional mystery shoppers to assess service levels according to the criteria identified. Call centers, for example, used to only count the number of abandoned calls or call duration as measures of customer service delivery, they now use instruments such as audiotaping phone conversations between service agents and customers and use coaches to randomly listen to conversations to assess the qualitative aspects of call handling. To enable us to consider different measurement instruments, I have outlined a number of different measurement instruments that can be used to measure performance:

- *Surveys and questionnaires* provide a relatively inexpensive way of collecting data on performance from a large pool of people who might be in different locations.[22] This can be done via mail, e-mail, Internet or phone. One big problem with this is that there has been a huge influx of surveys over the past few years, as more and more organizations require data for their nonfinancial indicators. The consequence of this is that it is getting harder to make people complete a survey. It is always a good idea to reduce the amount of time and effort required to collect performance data, not only for your organization, but also for your customers, employees, suppliers, etc. Intangibles such as employee engagement, corporate culture, customer attitudes, innovation climate or brand image are areas where surveys are regularly used as measurement instruments.
- *In-depth interviews* are guided conversations with people, rather than structured queries such as surveys. They involve putting forward open-ended (how, why, what) questions in a conversationally, friendly and nonthreatening manner.[23] Interviews can be conducted face-to-face or via telephone or videoconference. Interviews enable us to interact directly with respondents

and may result in new insights about performance. They provide examples, stories, and critical incidents that are helping us to understand performance more holistically.[24] In-depth interviews can, for example, be used to assess elements such as relationship with key customers, suppliers or partners. In addition to a performance score, it can also yield invaluable contextual information about, for example, how the relationships between key customers, partners or employees can be improved.

- *Focus groups* are facilitated group discussions (5–20 participants) in which participants can express and share their ideas, opinions and experiences. They provide a unique and interactive way to gather information and allow the collection of rich, qualitative information. Focus groups are good ways of assessing employee- and customer-related performance indicators such as customer experience, customer or staff engagement, team-working climate or trust.

- *Mystery shopping approaches* are the assessment of a service by a 'secret shopper' posing as a client or customer. Some organizations have in-house programs, whereby they employ their own mystery shoppers; other organizations hire external suppliers to provide this service. The beauty of this assessment approach is that it is less intrusive than surveys or interviews. Many government, public sector and not-for-profit organizations have started using mystery shopping to assess customer experiences. Trained mystery shoppers can also be used for many other internal performance assessments such as organizational culture or atmosphere.

- External assessments: External organizations and institutions can provide independent performance assessments and indicators. Independent surveys that measure the brand recognition, customer awareness or market share in specific segments are good examples of external assessments. An independent company creates a set of criteria and then measures everyone against these criteria to assess, for example, the relative position or values of brands or corporate reputations. The advantage of such external and independent assessments is that it provides data that allows comparisons between organizations. However, the problem with external assessment is that they might be too generic and often use assessment approaches that don't provide the answers to the internal KPQs. External assessments are best used as supplementary data to cross-check and validate other internal indicators.

- *Observations* allow us to collect information by observing situations or activities with little or no manipulation of the environment. The technical research field linked to observations is modern ethnography. Ethnography, which has its historic roots in the research of cultures and societies, uses observations and fieldwork to look at phenomena in a holistic manner. The method was founded on the idea that some phenomena cannot be accurately understood by only looking at some parts independently of each other. The observer can either take the role of a passive onlooker/outsider or can become involved in activities and, therefore, take the role of partial or full participant observer.

The power of using observation methods is that it engages all of our senses not just our sight. It enables us to take in and make sense of the entire experience through our nose (smell), eyes (sight), ears (hearing), mouth (taste), and body (touch). Unlike other data collection methods, observation data can provide us with a more holistic understanding of the phenomenon we're studying.[24]

Observation outputs can take the format of score sheets, checklists, narrative reports and video- or audiotaping. Observations have been successfully used in assessing organizational culture, skill and experience levels of employees, emotional intelligence and creativity. Another example is employee safety. Instead of waiting for accidents and injuries to occur and then count those, so-called safe behavior measures can be used: Observers proactively look for safe behaviors that would prevent the most common accidents and record those on a behavioral observation form. This information can then be shared and at the same time immediate feedback can be provided on potentially unsafe behavior. Ethnography techniques include direct, first-hand observation of behavior, including participant observation, and conversation with different levels of formality such as small talk or structured interviews. These observations can be recorded for a limited time or over a continuous long-term timeframe.

- *Peer-to-peer evaluation* is the assessment of performance in which participants vote or assess each other's performance. This can either be done openly or anonymously and enables people to learn from each other and to consider their own performance from the perspective of other people. Peer-to-peer evaluations have been successfully used to gauge elements including trust, knowledge and experience, teamwork and relationships.

There are many more fascinating ways of collecting qualitative performance data – for more information and examples see the *Handbook of Qualitative Research*.[25]

TRIANGULATE INDICATORS

It is a good idea to collect performance data using different techniques and methodologies. This allows organizations to contrast and compare the information gathered from different methods. This is called 'triangulation'. The rationale behind it is that the more information we have from as many possible sources – which all have advantages, disadvantages and different biases – the greater the likelihood that the information is reliable.

Organizations are often unaware of biases in their data collection. A frequently cited anecdote is the Wald story.[26] Abraham Wald was a statistician who helped the air force during World War II to assess where airplanes were most vulnerable to enemy fire. The plan was to subsequently reinforce the most vulnerable parts of the plane. Each airplane was examined for bullet holes and the areas that were disproportionately more often hit than others were identified. The air force thus concluded that the areas with the most bullet holes should be reinforced. However, Wald made them aware of the bias in the sample. Only airplanes that returned to the base were examined and included in the analysis. This, therefore, shows that the areas with many bullet holes have proved to sustain enemy fire and so these planes returned safely, whereas the areas with no bullet holes might be the best to reinforce since planes hit in these areas did not return.

Triangulation means that organizations collect data from different data sources (e.g. interviews with board members, middle managers, and frontline workers), use different methodologies (e.g. survey 70% and interview 30% of your suppliers) or use different people to conduct the data collection. This can reduce bias and increase reliability.

INDICES

When it comes to performance in government, public sector and not-for-profit organizations, it is rare that a single performance indicator will give us sufficient information. It can therefore be a good idea to create indices that combine different measures into one index. This allows organizations to get a more rounded and balanced view on their performance. Let me illustrate this point using human health. If your doctor came along and only measured the blood pressure to assess your health, then this would not be sufficient. However, by measuring blood pressure, cholesterol, body mass index and blood tests, together with a number of other tests, and combining these into a health index provide a much more balanced and reliable assessment of physical health. The same is true for organizations. If an organization wants to measure customer relationships, a number of indicators such as loyalty, trust, commitment, profitability, referrals, etc. can be measured and combined into a customer relationship index.

EXPERIMENTING WITH INNOVATIVE PERFORMANCE INDICATORS

In this section, I want to provide a number of examples that illustrate how some organizations have put the ideas and tools introduced in this chapter into practice and designed innovative, relevant and meaningful ways of assessing performance in their organization.

Case Study: Assessing Performance Improvement at the Orchestra[27]

Orchestra performances or theater shows would usually be classed as difficult to measure. Some would argue that the product is too intangible and the assessment of performance too subjective. However, orchestras or theater crews still need some indicators about their performance delivery and whether it is getting any better or not. Douglas Hubbard argues in his book on measurement that[28] 'For those things labeled "intangible", more advanced, sophisticated methods

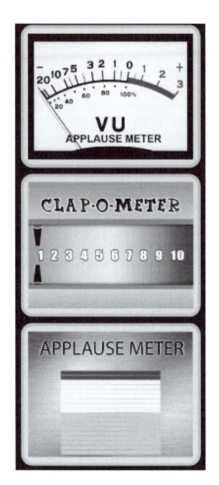

FIGURE 8.4   Applause meter.

are almost never lacking. Things that are thought to be intangible tend to be so uncertain that even the most basic measurement methods are likely to reduce some uncertainty'.

A good example comes from the Cleveland Orchestra,[27] who wanted to measure whether their performances were improving. They recently had a change in conductor and felt it would be important to gauge the performance impact and anticipated improvements over time. So the question they wanted to have an answer to was: How well is our performance improving? When they first thought about possible indicators, they considered customer satisfaction surveys. But they quickly ran into problems with this idea: Who do we ask? How can we create a representative sample? Will people actually want to complete surveys when they come to the performance? Is it better to do quick face-to-face surveys in the foyer or paper-based surveys? What kind of scale can we use? How will we know that our performance has improved if we don't survey patrons every night for some time? Initial thoughts were to create a rating scale from poor to excellent and then combine the answers into some sort of customer satisfaction index.

However, they realized that a survey would be costly, complicated and unreliable, and that it wouldn't really be helping them to answer their question. Instead, they started counting the number of standing ovations. They were aware that this was a relatively crude indicator but one that was giving them the information they needed. After a while of using this simple indicator, they felt that there were no obvious differences between performances where the number of standing ovations differed by just a couple, if they were able to see a significant increase over several performances then they could start to draw some useful conclusions. Douglas Hubbard said, 'It was measurement in every sense, a lot less effort than a survey, and – some would say – more meaningful. (I can't disagree.)'.

It is easy to see how this indicator can evolve into something slightly more sophisticated, such as measuring the length and the noise levels of the applause. Relatively inexpensive measurement instruments – so-called applause meters or clapometers – are now available to measure the noise level and length of applause (see Fig. 8.4). The results can then be recorded and mapped over time to see trends and performance improvements.

The Boston Symphony Orchestra went even further and wired 50 audience members with sensors during one of their concerts.[29] As part of a scientific experiment, members of the audience were wearing sensors on their arms and fingers, allowing their bodies to tell what kinds of emotional intensity they were feeling before, during and after the performance. While this might still be a vision for the future, it provides some food for thought about innovative ways of measuring performance.

Case Study: Indicators for Staff Engagement

Many organizations feel that their people with their skills and knowledge are their most important assets and the key enablers of future performance. In an organization, the management team had conducted some research and reached agreement that one of the key enablers of success was the level of staff engagement. According to the Gallup Organization *engaged employees* are passionate about what they do, they feel a strong connection to their company and perform at high levels every day while looking for ways to improve themselves and the company as a whole. *Not engaged employees* on the other hand show up every day and put in just enough effort to meet the basic requirements of their jobs. Without passion or innovation, these employees neither commit to the company's direction nor do they work against it. *Actively disengaged employees* are those who present a big problem for businesses. Negative by nature, these people are unhappy in their work and they compound their lack of productivity by sharing this unhappiness with those around them. According to Gallup Research, an average organization has about 25% engaged employees, just over half not engaged employees, and just under a fifth actively disengaged employees. Managers in this case study firm were keen to improve their ratio and ensured more employees were closely engaged.

In the past, this organization had conduced traditional staff satisfaction surveys but found that even though people might have indicated that they are satisfied with their jobs, many of them were not engaged. Managers in this organization believed that engagement is much more important than staff satisfaction as it indicates how passionate people feel about their jobs and how connected they feel to the organizations. Managers agreed to the following KPQ: 'To what extent are our employees engaged?' When they researched existing data collection methods they came across the Q12 survey tool[30] that has been developed by the Gallup Organization. This 12-question survey has been designed to assess engagement, especially on an emotional level. After some deliberation, the management team felt that this survey was right for them and would allow them to gain the information to answer their KPQ.

In addition, the use of this survey would allow the organization to benchmark itself with similar organizations. The following 12 questions, based on the Q12 survey, were incorporated into the staff survey at this organization:

1. Do I know what is expected of me at work?
2. Do I have the right materials and equipment I need to do my work right?
3. At work, do I have the opportunity to do what I do best every day?
4. In the last seven days, have I received recognition or praise for doing good work?
5. Does my supervisor or someone at work, seem to care about me as a person?
6. Is there someone at work who encourages my development?
7. At work, do my opinions seem to count?
8. Does the mission/purpose of my organization make me feel my job is important?
9. Are my coworkers committed to doing quality work?
10. Do I have a best friend at work?
11. In the last 6 months, has someone at work talked to me about my progress?
12. This last year, have I had opportunities at work to learn and grow?

The management team also decided to poll a representative sample of their employees every month so that they get regular checks of any possible changes. Each employee still receives their survey only once a year but the company gets valid data every month to answer their question and test the impact of staff engagement on retention, satisfaction and performance levels. The results of staff engagement are now reported to the senior management team on a monthly basis. The data is provided in aggregated form (staff engagement index) and compared with the positions of similar organizations. Engagement is best reflected by changes over time and the data is therefore presented in a trend chart together with a narrative commentary by the human resources director to put the performance data into context and to highlight the key issues and learning points.

Case Study: Assessing Stress Levels at a Charity

An international charity I was working with also felt they were not getting the information they needed from their annual staff satisfaction survey. For them, engagement wasn't the problem. In fact employees and volunteers in this organization were very engaged and passionate about the good cause. However, their problem was stress levels. Because people felt so passionate and committed to making a difference to the people they were working for, many worked too hard, which started to take their toll and sick leave went up. The management team felt that it needed to get a better handle on this and make sure they managed stressful periods much better to avoid burnouts and stress related sick leave.

The management team's KPQ was: To what extent are the stress levels getting outside the normal range. The management needed this information to make decisions about project scheduling to ensure overly stressful periods are avoided and workload levels are more evenly distributed. They found some research to show that shorter periods (3–4 days) of higher stress levels are okay and can even be motivating (e.g. the successful delivery of an important project) as long as they are followed by some less stressful periods. The management team considered surveys and questionnaires but felt they were too intrusive and 'in your face'. The feeling was that people wouldn't want to complete a survey at the frequency needed to get the necessary data. In order to capture whether people were stressed for longer than 3 days in a row meant that data was required at least every 2 days and surveys were just not feasible.

Instead, the organization introduced two different foot mats in the staff entrance of the building – one red and one green. These foot mats were fitted with sensors to detect which one people stepped on when they entered the building in the morning and when they left the building in the evening. Employees were asked to step on the red mat if they felt their stress levels were too high and on the green mat when their stress levels were okay. Each day, the data from the sensors were captured in a database and a summary report was e-mailed to the management team. This daily report enabled the management team to relate stress levels to specific projects and programs, and they were able to redistribute the workload to reduce stress levels. Again, the beauty of this system is that managers can try out different project combinations and get immediate feedback on the impact on stress levels. This process almost eliminated stress-related absenteeism and allowed the organization to get a real handle on an important issue. After using this method for about 6 months, the managers felt they were able to schedule and mange workloads much better. There was no need anymore for the daily reports to management, which were changed to weekly reports and later to monthly overviews. Today, the organization has stopped using the foot mats because the issue has been addressed and the necessary learning has taken place. It is just keeping an eye on stress-related absenteeism to make sure the issue is not coming back. The foot mats are now in storage and, if necessary, could be used again.

Case Study: Indicators for Corporate Reputation

A government organization wanted to better manage its perceived reputation. Similar to most government and not-for-profit organizations, a positive perceptual reputation is crucial. Managers in this organization felt that a positive perceived reputation has an impact on, for example:

- public opinion and how the press would treat that organization,
- potential future employees who will be attracted by a good reputation,
- central government that has to increasingly make choices about optimal spending and reputation is an important factor,
- this organization that has to rely on partnerships to deliver their outcomes, and to be successful in these partnerships the other organization have to respect this organization.

In the past, they used fragmented approaches and many ad hoc studies of how the different stakeholders perceived the organization. However, this was not painting a coherent picture and it was difficult to draw meaningful conclusions from these different studies – all of which were using different techniques and questions.

The management team responsible for corporate reputation did some research and found the work on the corporate reputation quotient (RQ) developed by Harris Interactive in association with the Reputation Institute.[31] RQ is a comprehensive measuring method of corporate reputation that was created specifically to capture the perceptions of the key corporate stakeholders such as consumers, employees or key influencers. However, while this tool was very appealing, it had been developed for corporate for-profit organizations and not all aspects of this RQ tools seemed relevant. The management team discovered further research, which customized the original six dimensions into seven dimensions for government and not-for-profit organizations.[32] The management team amended some elements of this framework slightly and started measuring the perceived corporate reputation using a survey that included the following questions accompanied by a 5-point Likert rating scale (see Fig. 8.5). This survey is now run every 6 months and

Describes the
organization …

1 …Very well   ◯

2 …Well   ◯

3 …Fair   ◯

4 …Poorly   ◯

5 …Very poorly   ◯

FIGURE 8.5   Point Likert scale to RQ.

has enabled the management team to get a much better handle on the management of the organizational reputation.

1. Emotional appeal
   - I have a good feeling about the organization.
   - I admire and respect this organization.
   - I trust this organization.
2. Services, products and tasks
   - The organization has a customer-oriented attitude.
   - The organization delivers appropriately high-quality services (within its financial means).
   - Innovation is important to this organization.
3. Vision and leadership
   - The organization has excellent leadership.
   - The organization has a clear vision for the future.
   - The organization is politically aware.
   - The organization is agile.
4. Workplace environment
   - The organization is well-managed.
   - The organization looks like a good organization to work for.
   - The organization looks like it has got good employees.
5. Social and environmental responsibility
   - The organization deals with matters that are important and relevant to public society.
   - The organization is an environmentally responsible organization.
   - This organization is ethical, honest and conscientious.
6. Financial performance
   - Financial means are used effectively and efficiently.
   - The organization is transparent about it resources and spending.
   - The organization is financially sound.
7. Communication
   - The objectives and tasks of the organization are clear to me.
   - The organization is transparent in its decision-making.
   - The organization engages with its stakeholders.

Case Study: Gauging Organizational Culture and Leadership

In an organization leadership was established as a critically important issue. The organization had just gone thought a merger of three government departments, which caused a lot of uncertainty and negative feelings. Strong leadership was required to ensure the organization would come together into one cohesive new organization. All senior managers and directors were sent on a customized leadership training. The organization now wanted to assess the impact of this training and see whether the leadership perception – which was at an all time low – was improving. Instead of surveys, interviews or focus groups, the organization came up with a very simple way of gauging the improvement in the leadership. In the staff canteen, they installed a tube in which people could put different colored balls. When members of the staff returned their food tray, they could pick up a red, white or green ball indicating that they had experienced good leadership (green), bad leadership (red) or neither good nor bad leadership (white). This ball was then placed into the tube, which went down one floor into a big collection container (see Fig. 8.6). This gave the management team a very good indication about the leadership perception and how it changed over the weeks. In this case, the number of red balls increased toward the top, indicating that the leadership got worse as time progressed. This was an important impetus to change the leadership training.

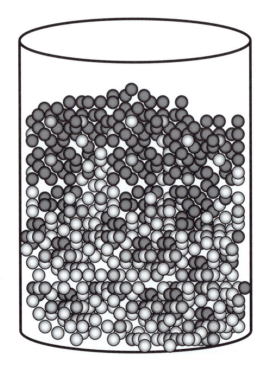

FIGURE 8.6   Measuring leadership.

Case Study: Measuring Customer Service Quality

A central government agency wanted to improve its service quality. A key strategic objective was to deliver more customer-centric and value-for-money services to its customers. For this it needed to understand the perceived service needs of its customers. After some research and deliberation that organization decided to use the SERVQAL method as their performance assessment approach. The SERVQAL method was developed by academics[33] to measure the gap between service quality performance and customer service needs. Originally it was measured on the following 10 aspects of service quality: reliability, responsiveness, competence, access, courtesy, communication, credibility, security, understanding or knowing the customer and tangibles. However, later this model was simplified and refined to contain the following aspects (representing the acronym RATER):

- Reliability
- Assurance
- Tangibles
- Empathy
- Responsiveness

This was used as an efficient model in helping this organization to shape its efforts in bridging the gap between perceived and expected service and led to much improved customer satisfaction scores.

After this organization had used the model for 3 years, it felt that the measurement efforts were too much and were not justified any more. The management team felt it had a much better handle on the service delivery and wanted to concentrate on measuring customer satisfaction in a more meaningful and simpler way. Again, some research led to the idea of just using one question to assess customer satisfaction: How likely is it that you would recommend (organization X or service Y) to a friend or colleague? Using this single question allows organizations to gain a 'Net Promoter Score' or NPS, which basically represents the percentage of customers whose answers identify them as promoters minus the percentage whose response indicates they are detractors. This idea is based on the work by Fred Reichheld, Director Emeritus and Fellow at Bain & Company, who divides customers of any organization into three categories:[34]

- Promoters (those who answer 9 or 10) are loyal enthusiasts who keep buying from a company and urge their friends to do the same.
- Passives (those who answer 7 or 8) are satisfied but unenthusiastic customers who can be easily wooed by the competition.
- Detractors (those who answer 0 through 6) are unhappy customers trapped in a bad relationship.

Using just the single question together with a 0–10 point rating scale with '0' representing the extreme negative and '10' representing the extreme positive end allowed this organization to categorize their customers according to their answers and gain a much better understanding about customer satisfaction (see Fig. 8.7 for the assessment scale used).[35]

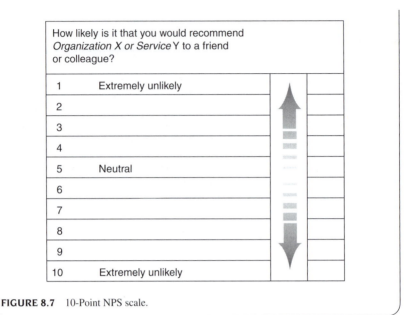

How likely is it that you would recommend *Organization X or Service* Y to a friend or colleague?		
1	Extremely unlikely	
2		
3		
4		
5	Neutral	
6		
7		
8		
9		
10	Extremely unlikely	

FIGURE 8.7   10-Point NPS scale.

Case Examples: Observations and Engaging Customers in Performance Assessment

Observations and engaging customers in the assessment of performance is not yet used too much in government, public sector and not-for-profit organizations. However, I feel that there are huge opportunities here. Some commercial organizations in particular have been able to engage their customers in the process of assessing performance. What they have done makes the measurement process interesting for the customers or provides them with some rewards for assessing performance.

One example comes from Thomas Cook, one of the world's leading leisure travel groups with sales of around £9 billion, 19 million customers, 30 000 employees, a fleet of 97 aircraft, a network of over 3000 owned or franchised travel stores and a number of hotels and resort properties. Thomas Cook started asking normal tourists who booked their holiday with them to be performance reviewers for them. This year they invited hundreds of customers to provide feedback on the different elements of their holiday. Each of these secret holiday reviewers is given a digital camera to capture the good and bad things they experience on holiday. As a little thank you gesture, the customers are able to keep the camera after the mission is completed. This is a really good example of how to engage customers and learn from their experiences. Customers actually really enjoy their 'secret mission' and tend to develop a closer and more loyal relationship with Thomas Cook as an organization. Instead of collecting the survey data, which is often seen as a hassle by customers, the review missions are seen as novel and exciting. The information gained from these review missions can then be used to better identify what really matters to customers. These insights can then used to brief 'professional mystery shoppers' to assess service levels according to the identified criteria.

Other companies such as BMW have created and sponsored user groups in which enthusiastic drivers of their cars happily share their experiences with the cars and the service levels provided by BMW. This provides invaluable insights into aspects of performance. I feel that organizations such as local governments and charities should have it relatively easy to engage people in performance assessment because what they do closely relates to people's lives (e.g. local government) or touches them on an emotional level (e.g. charities).

Another example comes from alcoholic drinks company Scottish & Newcastle (S&N), which has strong positions in 15 countries and which was recently acquired by a consortium of Carlsberg and Heineken. The work that I want to talk about here is specific to one of their leading beer brands (John Smiths), which is selling over 1 million pints every day. The company created a significant campaign called 'The Biggest Round'. Here, representatives employed by the experimental marketing agency BEcause would go into pubs and bars and observe the behavior of customers. In order to engage people, they would ask them 'Can I buy you a drink'? This then gives the representative an opportunity to talk to customers about what they like and don't like in a very casual and natural manner. These face-to-face conversations are now increasingly being used to gain important insights into customer behaviors and choices, which in turn is used to improve their marketing. Alison Nolan, Head of the S&N Account at BEcause, comments, 'Today's consumers quickly decide which brands they want to interact with and which they are going to ignore. Those that succeed in the future will be those who talk with consumers, rather than at them. That individual approach is the key to the ongoing success of this campaign'.[36]

I hope that the above examples have provided some ideas of how more innovative and methodical measurement approaches can lead to meaningful and relevant insights that help you answer the critical performance questions. I also hope they have shown that measurement doesn't need to be complicated and number focused.

SUMMARY

- A performance indicator provides us with evidence and information that help us to reduce uncertainty, answer our open questions and make better decisions.
- The concept of indicators is much broader than just the narrow link with numbers and counting; it includes methods such as written descriptions, observations, interview data, symbols, color codes, etc.
- It is completely normal and legitimate to use proxy and indirect performance indicators. We don't need 'perfect' indicators; instead we need appropriate accuracy based on our information needs and requirements.

- I have introduced a 10-step performance indicator decision framework which addresses the key questions and decisions someone has to make when designing relevant and meaningful performance indicators.
- I have outlined a comprehensive 20-element performance indicator design template. This design template is applied in conjunction with the performance indicator decision framework. It covers elements relating to indicator descriptions, data collection, targets, indicator quality and performance reporting. The template ensures that an indicator is designed compressively and that the relevant indicator information is captured in one place.
- Finally, I have provided a number of case studies to show how organizations have developed innovative indicators in relation to important areas such as service quality, staff engagement, corporate reputation, culture and leadership among others.

REFERENCES AND ENDNOTES

1. See for example: Hubbard, D. (2007). *How to Measure Anything: Finding the Value of Intangibles in Business*. Wiley Hoboken, New Jersey, p. 20.
2. Hubbard (2007), p. 31 (see note 1 above).
3. From: Asimov, I. (1965). *Of Time and Space and Other Things*. Lancer Books Inc., New York, NY.
4. Two sentences based on: Robinson, A. (2007). *The Story of Measurement*. Thames & Hudson, London, p. 19.
5. This section is based on: Whitelaw, I. (2007). A Measure of All Things – The History of Measurement Throughout the Ages. Quid Publishing, Hove.
6. See for example: Spitzer, D. R. (2007). *Transforming Performance Measurement: Rethinking the Way We Measure and Drive Organizational Success*. American Management Association, New York. Spitzer talks about appropriate accuracy (e.g. p. 207).
7. See for example: Hubbard (2007), p. 22 (see note 1 above).
8. See for example: Hubbard (2007), p. 14 (see note 1 above).
9. As reported in Neely, A., Kennerley, M. and Adams, C. (2002). *The Performance Prism*. FT Prentice Hall, London, p. 42.
10. The ideas and thinking I present in this section build on work done by, for example, Bourne, M., Neely, A., Mills, J., Platts, K. and Richards, H. (2002). *Getting the Measures of Your Business*. Cambridge University Press, Cambridge, p. 69; Adams, C., Kennerley, M. and Neely, A. (2002). *The Performance Prism: The Scorecard for Measuring and Managing Business Success*. FT Prentice Hall, London, p. 34; Bourne, M. C. S., Mills, J. F., Neely, A. D., Platts, K. W. W. and Richards, H. (1997). Designing Performance Measures: A Structured Approach. *International Journal of Operations & Production Management*, 17(11–12), 1131.
11. See for example: Preskill, H. and Russ-Eft, D. (2001). Evaluation in Organization: *A Systematic Approach to Enhancing Learning, Performance, and Change*. Perseus, Cambridge, MA, p. 178.
12. Likert, R. (1932). A Technique for the Measurement of Attitudes. *Archives of Psychology*, 140, 1–55.
13. See: *Target Setting – A Practical Guide*, IDeA/Audit Commission PMMI, http://www.idea.gov.uk/idk/aio/985665

14. See for example: Locke, E. A. and Latham, G. P. (1978). *A Theory of Goal Setting and Task Performance*. Prentice Hall, Eaglewood Cliffs, NJ; Latham, G. and Edwin, L. (2002). Building A Practically Useful Theory of Goal Setting and Task Motivation: A 35-Year Odyssey. *American Psychologist*, 57(9), 705–717.

15. Based (but amended and extended) on: Target Setting – A Practical Guide, IDeA/Audit Commission PMMI, http://www.idea.gov.uk/idk/aio/985665.

16. Examples based on the material developed by IDeA/Audit Commission PMMI Project (*Performance Management, Measurement and Information*), http://www.idea.gov.uk/idk/core/page.do?pageId=845670 and *On Target – the Practice of Performance Indicators*, a management paper published by the Audit Commission.

17. Austin, R. D. (1996). *Measuring and Managing Performance in Organizations*, p. 66, p. 193. Dorset House Publishing, New York.

18. Gray, D. J. (2005). A Multi-Method Investigation into the Costs and into the Benefits of Measuring Intellectual Capital Assets (unpublished Ph.D. thesis). Cranfield School of Management, Cranfield.

19. See for example: Spitzer (2007), p. 85 (see note 6 above).

20. See for example: Ketokivi, M. A. and Schroeder, R. G. (2004). Perceptional Measures of Performance: Fact of Fiction. *Journal of Operations Management*, 22(3), 247–264; Boyd, B. K., Dess, G. G. and Rasheed, A. M. A. (1993). Divergence Between Archival and Perceptional Measures of the Environment: Causes and Consequences. *Academy of Management Review*, 18(2), 204–226; Venkatraman, N. and Ramanujam, V. (1987). Measurement of Business Economic Performance: An Examination of Method Convergence. *Journal of Management*, 13(1), 109–112.

21. Spitzer, Dean R. (2007). *Transforming Performance Measurement: Rethinking the Way we Measure and Drive Organizational Success*. American Management Association, New York.

22. For more information see for example: Dillman, D. A. (1999). *Mail and Internet Surveys: The Tailored Design Method*. Wiley, New York.

23. Yin, K. (2003). Case Study Research Design and Methods (Applied Social Research methods series, Vol.5). Sage, Newbury Park, CA.

24. Russ-Eft, D. and Preskill, H. (2001). *Evaluation in Organization – A Systematic Approach to Enhancing Learning, Performance, and Change*. Perseus, Cambridge, MA.

25. For example: Denzin, N. K. and Lincoln, Y. S. (ed.) (2005). *The Sage Handbook of Qualitative Research*, (3rd edn.). Sage, Thousand Oaks.

26. See for example: Mangel, M. and Samaniego, F. J. (1984). Abraham Wald's Work on Aircraft Survivability. *Journal of American Statistical Association*, No. 79, 259–267.

27. This example was outlined in: Hubbard (2007), p. 32 (see note 1 above).

28. Hubbard (2007), p. 33 (see note 1 above).

29. Elton, C. (2006). Measuring Emotion at the Symphony. *The Boston Globe*, April 5.

30. The Q12 survey tool was developed by the Gallup Organization (http://www.gallup.com); other staff engagement surveys are offered, for example, by Mercer and Satmetix Systems called Employee Acid Test and Employee Commitment Assessment.

31. The Reputation Institute is a research organization dedicated to advancing knowledge about corporate reputations. The Institute is sponsored by PricewaterhouseCoopers and Shandwick. Harris Interactive is a leading market research and polling firm. For more information please see: http://www.reputationquotient.com

32. The 7-dimension RQ for government and not-for-profit organizations is based on the work by: van Driel, O. R. (2002). Towards a Profitable Reputation Quotient for Governmental and Not-For-Profit Organizations. Corporate Communication Centre, Dallas. However, some elements of the survey have been amended.

33. Zeithaml, Parasuraman and Berry, (1990). *Delivering Quality Service: Balancing Customer Perceptions and Expectations*. Free Press, New York.

34. See for example: Reichheld, F. (2006). *The Ultimate Question: Driving Good Profits and True Growth*. Harvard Business School Press, Boston, MA; Reichheld, F. (2003). The One Number You Need to Grow. *Harvard Business Review*, December.

35. Measuring Satisfaction on a 0–10 Scale: A Satmetrix Systems White Paper, http://www. netpromoter.com/pdfs/0-to-10-Final.pdf

36. Source: http://www.creativematch.co.uk/viewnews/?94103 (July 2008).

Part III

Learning and Improving Performance

So far in this book, we have looked at how we identify and agree what matters and how we collect the right management information in our organizations. This is good as far as it goes; however, unless we turn the management information into insights that help us deliver on our strategy and make better decisions that improve performance, all efforts so far are in vain. There are too many government, public sector and not-for-profit organizations that believe by just measuring performance they will somehow improve. This is not the case! As part of any attempt to measure and manage performance, we need to create the right organizational context and the appropriate processes that help us turn our information into meaningful insights and learning.

Only if organizations address this final piece of the performance management jigsaw can they expect to see any real differences in performance. In this part of the book, I look at how we can make the best use of the management information to learn and improve performance. In Chapter 9, I discuss one of the most crucial elements of good performance management, namely, how to create a performance-driven culture. Some of the key elements of a performance-driven culture are the appropriate leadership approach, the right reward and recognition system, appropriate reporting and communication processes and the right performance review routines.

In Chapter 10, I discuss how we can leverage performance management software applications to help us bring performance management to life. Performance management software applications can help us facilitate and automate many aspects of performance management and, most importantly, can help us engage everybody in the process. As part of Chapter 10, I also provide a framework for selecting appropriate software applications, together with a list of credible vendors.

In Chapter 11, I take a look at the current state of practice and what we can learn from it. Based on the world's largest and most comprehensive study of government and public sector performance management, I outline 10 principles of good performance management. See Fig. P3.1 for an overview of the chapters in this third and final part of this book.

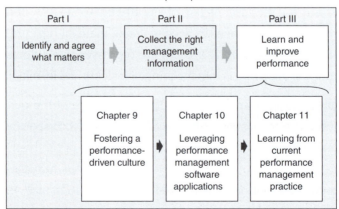

FIGURE P3.1   Overview.

Chapter 9

Fostering a Performance-Driven Culture

In order to successfully manage performance, organizations need to create the appropriate organizational culture. We have to move away from the 'command-and-control mentality' or the 'reporting-only mind-set' of measurement, in which backward-looking, pseudo-relevant metrics are collected, reported or being used to punish people. In Chapter 6, I have discussed the dangerous side effects of the command-and-control and reporting-only models. Instead, here I want to discuss ways of creating the right performance-driven culture in which an enabled learning environment allows us to use indicators to learn, challenge and improve future performance.

The questions I address in this chapter include:

- What is a performance-driven culture?
- How can we create the right performance-driven culture?
- How can we create performance-driven leadership?
- How can we appropriately link performance with rewards and recognition?
- How can we better report performance?
- How can we make sure performance is reviewed appropriately?

People and the emphasis on learning and improvement are at the center of a performance-driven culture. Dean Spitzer, performance measurement thought leader at IBM, makes an important point when he says:

> It is people who will ultimately determine the effectiveness of the measurement system, because measurement data is of no value without human involvement. It is human beings, not machines, who turn data into information, information into insights, insights into knowledge, and knowledge into wisdom.[1]

We therefore need to make sure these human beings operate in the right environment and with the right mind-set to ensure learning and performance improvement takes place.

Every organization has got a culture. If the culture is right, it can be one of the most powerful success factors in any performance management initiative. However, the culture can also be an important inhibitor for successful performance management. I believe we need to spend a bit more time on making sure we create the right atmosphere and behaviors in our organizations. In his article on performance-driven culture, Philip Atkinson writes that 80% of organizational culture exists by accident or default, rather than design.[2] I am not suggesting that you run a big culture diagnostic and then start a big culture change program. Instead, I want to outline the key tools, techniques and behaviors that will contribute to a performance-driven culture. I believe that creating a performance-driven culture is easier than most people think and I hope I can help to demystify the process. Before we discuss the key building blocks and tools of a performance-driven culture, let me first outline what I mean by a performance-driven culture.

WHAT IS A PERFORMANCE-DRIVEN CULTURE?

An organizational culture represents the shared underlying beliefs, norms, values, assumptions and expectations, which influence the typical patterns of behavior and performance that characterize an organization. The organizational culture influences the way things get done in an organization and therefore also governs the way people react to performance indicators and use performance information. Research has shown that creating a culture in which performance is recognized as a priority can have a significant and tangible impact on success.[3] One of the best descriptions of a so-called performance-driven culture that I have seen so far comes from an article on the topic written by Howard Risher, a pay-for-performance expert, published in The Public Manager:

> In an organization with a strong performance culture, employees know what they are expected to accomplish and are emotionally committed to organizational success. They believe in the mission and goals and are quick to put their energy into a task without being asked or monitored. Informal conversations with coworkers frequently focus on performance problems and recent organization results. They tend to celebrate successes as a team or group. The commitment to performance is a way of life in the organization.[4]

For me, a performance-driven culture means that people in an organization are continuously striving to learn and improve. At the center of a performance-driven culture is organizational learning and improvement (see Fig. 9.1). Enablers of a performance-driven culture are:

- a sense of community and a common purpose, which binds people together and provides the emotional commitment to success,
- clear and accepted accountability for results and performance across the organization, which provides responsibly and ownership,
- honesty and truth about performance results, which in turn creates an atmosphere of trust and mutual respect,
- a clear definition of what a performance-driven culture is, which creates an understanding and acceptance of a performance-driven culture throughout the organization.

The key tools or building blocks that will help to create a performance-driven culture are:

- a strong performance-driven leadership throughout the entire organization,

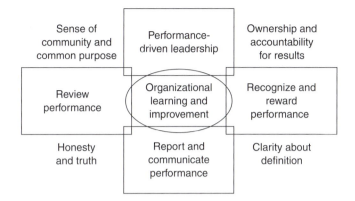

FIGURE 9.1 Performance-driven culture.

- a reward and recognition system that allows us to celebrate and recognize good performance,
- appropriate reporting and communication of performance information,
- suitable interactive performance review processes that engage people in a dialogue about performance and lead to learning, decision making and performance improvements.

Figure 9.1 shows the model of a performance-driven culture. It depicts how organizational learning and improvement sit at the center of the model, while the four building blocks or tools are positioned around the center. All of them overlap with each other and link to organizational learning and improvement. The four enablers of a performance-driven culture frame and complete the model.

ORGANIZATIONAL LEARNING AND A PERFORMANCE-DRIVEN CULTURE

Organizational learning is at the center of a performance-driven culture. Most theorists agree that organizational learning takes place when individuals and teams engage in a dialogue, reflection, asking questions, and identifying and challenging values, beliefs and assumptions.[5] I suggest that organizations create an enabled learning environment. An enabled learning environment is an organizational environment in which all employees are actively seeking new strategic insights, which are based on their understanding of strategy, key performance questions and the performance indicators collected, to allow them to challenge strategic assumptions, to refine strategic thinking, to learn and to make better decisions to improve future performance. The word 'enabled' stresses the fact that employees are also enabled or empowered to use strategic insights. Having insights about how to improve things without the authority to do something about it is often a source of employee frustration. In an enabled

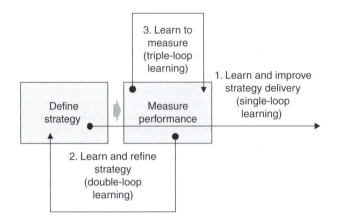

FIGURE 9.2   Performance management learning cycles.

learning environment, the value creation map, the key performance questions and the performance indicators become the means for providing information for learning, decision making and action.

Chris Argyris, professor at Harvard Business School, defines learning as occurring under two conditions. First, when an organization achieves what it intended to achieve and there is a match between the intention and outcome. Second, when an organization identifies a mismatch between the intention and outcome and this mismatch is corrected, so that a mismatch is turned into a match.[6] In order to learn, organizations therefore require an understanding of their intentions and a way to test the match or mismatch between their intended and actual performance. The value creation map and the value creation narrative make the organizational intentions explicit. They represent the assumed model of how the organization is intending to create outcomes and value. The performance indicators enable organizations to then test these assumptions. This allows individuals in the organization to reflect on the assumptions, learn from the insights and improve their decision making.

When it comes to performance management, it is possible to distinguish between three types of organizational learning: single-loop learning, double-loop learning and triple-loop learning (see also Fig. 9.2):[7]

- Single-loop learning: A thermostat is often described as a single-loop learner. A thermostat is programmed to detect states of 'too cold' or 'too warm' and then correct the situation by turning the heat on or off. Single-loop learning takes place in organizations when, for example, an agreed strategic objective gets implemented and measures are used to check any variance from the expected results. Here, measures are used to understand whether the objective was delivered and the information is used by managers to understand whether corrective actions are necessary.

- Double-loop learning: Double-loop learning takes place when, for example, an agreed strategic objective gets implemented and the insights from the measurement lead to questioning of the underlying assumptions and possible revision of the strategic objective or the value creation model. If a thermostat were able to ask itself such questions as 'why the thermostat was set to, for example, 68°F', or 'why it was programmed as it was', then it would be a double-loop learner.[6] During a double-loop learning process, the information is used to challenge strategy and its components in order to revise or refine the strategy.
- Triple-loop learning: In the triple-loop learning process, the organizations assesses its ability to assess performance with the given performance indicators. It evaluates how good the performance indicators and key performance questions are in helping the organization to improve and learn. If a thermostat were able to ask itself such question as 'is there a better way to assess temperature than using the inbuilt thermometer', then it would be a triple-loop learner.

During the single-loop learning process, the value creation model is taken as a given. Indicators are collected, analyzed and interpreted in order to take actions. Single-loop learning takes place when organizations review their performance against targets or intentions. Double-loop learning follows the same logic, but instead of only testing performance against intentions, the underpinning value creation model and its underlying assumptions are challenged. In the triple-loop learning process, we challenge our ability to assess the performance appropriately. In order to learn, it is important that our assumptions about the value creation model and our assessment capabilities, which often manifest themselves in the taken-for-granted behavior and opinions, are continually questioned, tested and validated.[8] In their book on balanced scorecard, Bob Kaplan and Dave Norton write:

> Of course, managers need feedback about whether their planned strategy is being executed according to plan – the single-loop learning process. But even more important, they need feedback about whether the planned strategy remains a viable and successful strategy – the double-loop learning process.[9]

I go even further than that and say that we need single-, double- and triple-loop reflections in our organizations and that it is not just the managers who need to question the underlying assumptions; it is everyone in the organization.

In order for any learning to take place, organizations need to create the right social context or culture. What I call a performance-driven culture is diametrically opposed to the traditional command-and-control environment and the reporting-only mind-set. In Chapter 6, I have outlined the limitations of measurement as a reason why the command-and-control environment will create dysfunctional behavior. There is another important reason why the command-and-control environment is no longer appropriate in today's business environment: because it inhibits learning.

CREATING A PERFORMANCE-DRIVEN CULTURE

Creating a performance-driven culture is not about esoteric communities in which everyone happily does whatever they want. It is about creating an environment in which performance is a priority and where trust, self-directed learning, mutual respect and support lead to personal commitment to continuous performance improvement.

Most people understand that ever-tighter budgets and tougher operating environments for government, public sector and not-for-profit organization will require more effective learning, broader empowerment and greater performance commitment from everyone in the organization.[9] In order to make learning more effective, organizations need to create a performance-driven culture. The problem is that there is no easy button we can press to switch on such a culture. The process will take time and efforts and is never likely to be complete. Having said this, there are a number of tools and building blocks we can put in place to foster and shape a performance-driven culture.

Where to start is a little bit of a 'chicken and egg' question – it's hard to say which must come first. Creating an enabled learning environment, addressing the four enablers and implementing the four building blocks of a performance-driven culture will help to change the routines and practices, and changing the way we do things has an impact on the perceived culture. In the following sections, I discuss the organizational learning element before I briefly discuss the four enablers and provide guidance on how to implement the four key building blocks of a performance-driven culture.

EMPHASIZING LEARNING AND IMPROVEMENT IS KEY

Emphasizing learning and improvement is the single most important aspect of creating a performance-driven culture. The National Performance Review report by the US vice president rightly states:

> Performance measurement systems should be about learning and improvement, they should be positive, not punitive. The most successful performance measurement systems are not "gotcha" systems, but learning systems that help the organization identify what works and what does not so as to continue with and improve on what is working and repair or replace what is not working. Performance measurement is a tool that lets the organization track progress and direction toward strategic goals and objectives.[10]

Too many organizations still operate the punitive command-and-control model creating a blame culture, which brings out the worst in people. Instead of a positive focus on learning, it creates fear, distrust, self-centeredness and protectionism. In such an environment, people are not willing to openly and voluntarily share their insights and knowledge. There is no incentive for collaboratively exploring performance improvement; therefore, real improvements and innovations are rare.

There is an interesting parallel that we can find in the education world, where different forms of assessing pupils or students yield very different outcomes. Traditionally, we tend to use *summative assessments* in schools. Summative assessment is an assessment, typically an exam or test, that determines the learning outcome of an academic program, let's say a language course, at the end of the program or at the end of a particular phase of the program. Such assessments are judgments about the student's learning, mostly in the form of a grade, which is given compared to some standard or to the performance of others. These assessments often have high stakes attached to them, for example, a qualification and access to university. Most exams and standardized tests today are summative in nature. They are seen to provide reliable and comparative data, and the assumption is that such tests produce improvements in student learning. However, this assumption is questioned by many since these assessments are not designed to provide contextualized feedback that is useful for helping students and teachers during the course of a program to improve learning.

By contrast, *formative assessment* is a feedback process into an ongoing program in order to improve the learning. It occurs when teachers feed performance information back to students in ways that enable the student to learn and fine-tune or modify what they have been doing, or when students can engage in a similar, self-reflective process.

Whereas in summative assessments the result (e.g. grade) is at the center of attention, in formative assessments the improvement of learning is the key objective. The former is backward looking, whereas the latter is about positively impacting on the future. Formative assessment is more about detecting learning shortcomings early enough and doing something about them. It also engages the students and provides them with useful information about their progress and any learning gaps, which they can then use to make decisions about how to improve future learning.

Research in this area provides strong evidence that formative assessment is a powerful means to improve student learning, whereas summative assessments such as standardized exams can, in fact, have a harmful effect.[11] An article on the topic highlights the fact that most classroom testing encourages rote and superficial learning.[12] Professors Paul Black and Dylan William found that teachers often emphasize quantity of work over high quality. Actual assessment practices show that marking and grading are overemphasized, while giving useful advice is underemphasized. Overall, summative assessments tend to have a negative effect on student learning. One expert on the topic talks about 'education by numbers' and the 'tyranny of testing'.[13]

This problem is made worse by the fact that, in many countries, schools or universities are now being assessed on the outcome of such standardized summative assessments. The laudable aim is to make schools accountable for their teaching quality and the progress in learning achieved by the students. The numerical outcomes of these assessments are then used to create, for example, school league tables, which are published in order to inform parents and students about the performance of different schools.[14]

The problem is that what is being measured is a proxy measure for learning that measures only the numerical outcome of the exams and not whether any real learning has taken place. This is made worse by the fact that the stakes are high not only for pupils but also for schools. In many countries, the league table results have an impact not only on reputation but also on funding. This mechanistic focus on proxy outcome measures leaves the system open to cheating and therefore can create dysfunctional behavior. Teachers might only teach what is important to pass the exams with little actual learning, and students might try to do as little as they can get away with to meet the minimum requirement. A number of recent reports have shown that many schools now overemphasize exam preparation and in one case the teacher spends two-thirds of the class time on mock exams. Suddenly, the emphasis is not on learning but on playing the numbers game.

The key reason for this dysfunctional system is the wrong approach toward performance management and a missing performance culture. Summative assessments focus on the past performance and provide little or no guidance on what could be done differently in future learning. Students who receive a grade at the end of a course can't do anything differently to improve it. At the same time, schools that are being assessed with a league table score receive no constructive feedback on how teaching quality could be improved. In short, this fuels the command-and-control or reporting-only mind-set and creates a massive loss of real learning potential.

Once we've been 'educated', we get a job (hopefully). In organizations we then tend to get much the same dysfunctional behavior and gaming of numbers. If we don't collect and apply the most relevant indicators (instead of opting for the ones that are easy to measure or provide a view only of historic performance) and if we don't create an environment in which indicators are used to inform our decision making and learning, then we are heading down the very same shallow track.

FOUR ENABLERS OF A PERFORMANCE-DRIVEN CULTURE

I have identified four important and interconnected enablers of a performance-driven culture. The enablers are not tools as such; they are more conditions that are necessary to make a performance-driven culture a reality. The tools, which are discussed later, can be used to enforce the enablers. Let's briefly discuss each of the four enablers in turn before moving on to the tools:

- Create a sense of community and common purpose: In today's world people want to know how they are helping to achieve the greater organizational goal. The commitment to serving the public and the delivery of a good cause are strong motivators for people in government, public sector and not-for-profit organizations. People also want to buy into the greater organizational goals and therefore create an emotional bond, which is expressed in a sense of community and common purpose. Professor Charles Ehin, of the Gore School of

Business, identifies sense of community and common purpose as important tenets for fostering positive social connections, voluntary collaboration and learning.[12] *Sense of community* gives individuals a sense of belonging, which fosters commitment, collaboration and mutual respect. A community is a social entity that serves both its members individually and the community as a whole. Underpinned by effective line-of-sight relationships, it creates a base for a group of people with shared interests, where compassion, empathy and trust pervade.[15] *Common purpose* is about meaning and an implicit agreement about direction. Can individuals associate themselves with the aspirations and perspectives of the organization or the team? Both, common purpose and a sense of community can often be linked to the boundary conditions of an organization (see Chapter 1) and a clearly articulated and agreed strategy (see Part I of this book). Genuine commitment to the performance of an organization simply does not happen unless individuals feel that they are empowered and respected partners on a joint journey.

- Insist on honesty and truth: Without honesty and truth, there can be no trust, and without trust, there can be no performance-driven culture. Professor Onora O'Neil makes a notable point in the 2002 BBC Reith Lecture when she argues:

> Perhaps the culture of accountability that we are relentlessly building for ourselves actually damages trust rather than supporting it. Plants don't flourish when we pull them up too often to check how their roots are growing: political, institutional and professional life too may not flourish if we constantly up-root it to demonstrate that everything is transparent and trustworthy.[15]

Insisting on honesty and truth means that everyone should admit mistakes and underperformance. Too many organization have what I call a 'red is bad culture'. The reason for this is more personal and often related to our human nature. In their professional environment, people are not very good at admitting failure and are, therefore, intrinsically unable to learn from their mistakes. We seem to have universal human tendencies to avoid embarrassment or threat, and we don't like feeling vulnerable or incompetent. In his *Harvard Business Review* article entitled 'Teaching smart people how to learn', Chris Argyris explains that failure produces defensive reasoning which can block learning, even if people's commitment to learning is high. He says:

> Put simply, because many professionals are almost always successful at what they do, they rarely experience failure. And because they have rarely failed, they have never learned how to learn from failure. So whenever their single-loop learning strategies go wrong, they become defensive, screen out criticism, and put the 'blame' on anyone and everyone but themselves. In short, their ability to learn shuts down precisely at the moment they need it the most.[16]

In order to create a performance-driven culture that has learning at its center, we need to overcome these barriers and see underperformance as an opportunity to improve.

- Ensure ownership and accountability: In an organization with a performance-driven culture, individuals and groups of people take ownership for the delivery of performance results and feel accountable for the achievement of the results. In such an environment, responsibilities are clearly assigned and well understood. A way to achieve clarity about responsibilities is by cascading the organization-wide strategy with its objectives down into the organization in order to create departmental and business unit objectives. These objectives can then be further broken down into individual performance plans with clearly articulated ownership and accountability.

- Clear definition of a performance-driven culture: Finally, it is important that people understand what a performance-driven culture is and what it entails. This can be used not only to communicate to existing employees but also to candidates and potential future employees. The right performance-driven culture will encourage your top-performing individuals to stay and other high-performing individuals will want to join. One way to explain your culture is to post it on the Internet. A good example is Weyerhaeuser, a major forestry company, which today owns or manages 21.5 million acres of timberland with offices or operations in 18 countries.[17] Their Web site states:

> Critical to building a culture of personal growth and engaging talented people is a disciplined performance management system. We are committed to cultivating a performance-driven culture that rewards results. That's why we have a rigorous performance management process, as well as a goal-setting process at all levels of the company. From our CEO to our business segments and across the corporate functions, we display our performance on critical measures through dashboards. Using a three-point scale of exceeds, achieves or below, we rate our performance in key areas. This goal-setting activity aligns team, department and individual goals to company goals. Progress is formally evaluated at mid-year and year end. […] Employees complete an annual performance management plan as described above, including specific goals relating to economic, social and environmental performance as appropriate. Similar goals are established for each business unit and for the company as a whole. Employee compensation is tied to the performance of the company, the business unit and the individual employee against these goals.[18]

Their employee climate survey has revealed that 90% of their employees understand what's expected of them on the job, 65% said people were held accountable to achieve their goals and 66% said they received regular feedback on their performance.[19]

THE FOUR BUILDING BLOCKS OF A PERFORMANCE-DRIVEN CULTURE

The four building blocks of a performance-driven culture are all tools that any government, public sector or not-for-profit organization can implement to create the right performance culture. Similar to the four enablers, the four building blocks are also interdependent and will achieve the best results if they are implemented together.

PROVIDE PERFORMANCE-DRIVEN LEADERSHIP

Dean Spitzer of IBM believes that leadership is the single most important aspect of transforming organizational performance measurement.[20] I agree that leadership sets the tone for everything, including performance measurement and performance management. Only if leaders in the organization champion a performance-driven culture and lead by example, can we successfully implement such a culture. It is a little bit like raising kids – it's not what you say; it's what you do that gets copied. People in organizations are 'boss watchers'[2] – they pay attention to what their leaders focus their attention on and copy the behavior their leaders display.

Buy-in and commitment to performance management among the executives and directors are vital, as this sends the right messages. However, it is important to stress that leadership does not reside solely among directors and senior managers. Leadership and the right behavior have to be demonstrated across all hierarchies. Let me outline some key behaviors senior leaders in an organization need to display in order to foster a performance-driven culture are:

- Showing visible commitment to performance management: Leaders must show that they value performance measurement, performance management and performance improvement. They send important signals just by talking about it and writing about the importance of managing performance. Actively and visibly supporting the other three building blocks of a performance-driven culture are other signals.

- Explaining the role of performance management: Leaders across the organization need to explain why performance measurement and performance management are important to the organization. They need to explain the key benefit the organization and individuals can gain from performance management. Experience has shown that this process of explaining has to happen repeatedly. In particular, executives and senior leaders must keep beating the drums about performance management as a tool to learn and improve.

- Getting actively involved in performance management: Visible and active involvement by senior executives and managers is a necessary part of successful performance measurement and management. Senior managers and directors should be actively involved in both the creation and the implementation of their organization's systems.[21] Top-level executives should not only personally articulate the strategy of the organization but also get actively involved in recognizing, reporting and reviewing performance at different levels of the organization.

- Moving from inspector to supporter: Leaders throughout the organization need to live the performance-driven culture by example. One of the best things they can do is to move from an inspector role to a supporter role. Leaders must make a commitment to help their people improve. Instead of inspecting performance, they need to provide guidance, coaching and advice on how performance can be improved.

I have seen an example where an organization successfully created a performance-driven culture based on learning, with mutual trust and excellent performance information. They built an environment where lots of information was coming out, where this information was openly shared using systems and wall charts and where people were looking at this information and gaining new insights and making better decisions. Until one day, a senior director walked in, looked at the performance results displayed on the wall charts, made some notes and, based on that information and without speaking to anybody, decided to move team members around and make two individuals redundant. Immediately after that event, all the performance charts came off the wall and people stopped sharing any performance information.[22] Creating the right leadership means being able to manage the fine line between learning and controlling. Without the commitment of managers and the right leadership, it will be impossible to create a performance-driven culture and to implement the other three building blocks of a performance-driven culture. Leadership is the foundation-building block.

REWARD AND RECOGNIZE PERFORMANCE

In order for performance management to be taken seriously, performance must have consequences. Both good and bad performance must have some sort of effect. If not, people will realize that performance measurement and performance management initiatives are not really important and have no real meaning. I see too many government, public sector and not-for-profit organizations that don't take any actions no matter how good or bad people perform. And those are the organizations where managers wonder why performance management is not really embedded in the organization.

Many experts advocate linking performance indicators to the pay of employees. While in theory this seems a good idea, in practice it is riddled with complications and dangers and rarely works well. Also, while private corporations have the freedom to set their payment systems in a way they want, most government and public sector organizations don't have this kind of freedom. As a consequence, very few government, public sector or not-for-profit organizations link their pay to performance. When the US Office of Management and Budget (OMB) announced The President's Management Agenda and kicked off its aggressive strategy for improving performance, it received a massive backlash and resistance from employees, which clearly shows that the culture is not ready to accept pay for performance.[4]

Implementing pay for performance in government, public sector and not-for-profit organizations may be difficult or impossible; however, every organization has a reasonably long list of ways that employees can be recognized and rewarded. Reward and recognition is more than money. Rewards can be organized as extrinsic or intrinsic. Extrinsic rewards are external to the person, for example, pay. Intrinsic rewards are internal to the person, for example, satisfaction or a feeling of accomplishment. Extrinsic rewards include financial and nonfinancial incentives.

It is important to understand that there are many ways to recognize and reward people without spending anything. Those nonfinancial reward mechanisms are usually available to government, public sector and not-for-profit organizations and should be considered. Research has also shown that nonfinancial rewards can be even more effective and powerful than monetary rewards. Here are some ideas from other government and not-for-profit organizations:

- A senior executive in a federal government agency regularly writes and sends handwritten personal notes to employees who deserve recognition.
- The chief executive of a local government body gave everybody a day extra holiday (time off) after they achieved a momentous performance turnaround.
- The chief executive of a major charity regularly sends a bouquet of flowers to the employees who have demonstrated performance-driven behavior.
- A government agency regularly holds 'performance parties' with free coffee and cookies for the departments or teams who performed well.

I believe the most powerful recognition and reward is the one we tend to forget too often: to say 'thank you'! Please don't underestimate the power of a 'thank you'. If said in earnest by managers or senior leaders, a simple 'thank you' can clearly outstrip the impact of a pay rise.

When it comes to rewarding performance, we have to be careful. Research firm Aberdeen Group finds in one of its studies that lots of lip service is paid to performance-driven culture. While organizations believe they have one, the facts show that the majority (73%) measures workers' success based on the successful completion of tasks but not on the quality of the tasks or whether this has helped to improve the performance of the organization.[23]

Here I have compiled a few tips of how to reward and recognize people in order to create a performance-driven culture:

- Celebrate success: Most government, public sector and not-for-profit organizations I have ever worked with are not very good at celebrating success. We need to do this much more often!
- Reward effort, not just success: When we reward and recognize people, we don't have to wait until a major outcome objective has been reached. We can reward them for the right efforts.
- Reward straight away: Studies show that if a person receives a reward immediately after they have done something well, then the effect is greater. The effect decreases as the duration between performance and reward lengthens. If you can, don't delay the reward.
- Don't create habits: If we regularly reward similar behavior, the rewards become a habit and lose their power. We need to avoid routine-like rewards that people will just take for granted.

- Don't 'hardwire' measures with compensation system: A big mistake is to link the performance of a number of specific measures with rewards. This mechanistic link based on proxies can drive many dysfunctional behaviors.
- Balance rewards for individual and corporate performance: We need to avoid rewarding only individual performance, as this can lead to increased competition and decreased team work. We need to balance individual performance with the performance of the teams, groups, departments and corporate organizations. I like to see a three-way split as individual performance, departmental or group performance and corporate performance.
- Use the whole spectrum of rewards and recognitions: There are other rewards than financial ones – we need to use these more often and start saying 'thank you'.

Linking reward and recognition to performance sends a clear and unambiguous message to the organization that performance management and performance improvement matter.

REPORT AND COMMUNICATE PERFORMANCE

Effective internal and external reporting and communication are important keys to successful performance measurement. However, the way in which most organizations communicate and report performance is not conducive to learning and is not helping to create a performance-driven culture. Most government, public sector and not-for-profit organizations today seem to have a tendency to produce cryptic spreadsheets containing performance data, which is then distributed as e-mail attachments. The fact that most people will have only one quick glance at the data and then quickly decide that they can't really make sense of it is rarely taken into account. If organizations are unable to engage people in a dialogue and make them reflect on performance, no learning will ever take place.

Effective communication with employees, customers, stakeholders and the public is vital to government, public sector and not-for-profit organizations today. In the end, they are the ones who judge whether the organization has been successful. In order to achieve effective communication, we need to make performance real to people; we need to make it visible and digestible.

Performance indicators are rarely reported in a manner that gives people sufficient information about the indicator and the performance levels. Any ambiguity leads to doubts, which in turn hamper understanding, decision making and learning. It is therefore important to provide a comprehensive picture of an indicator, and it is critical to bring across the message the indicator data is sending.

Best Practice Performance Reporting and Communication

The performance indicator design template outlined in Chapter 8 provides much of the information needed to explain to any person who receives the performance information what the indicator is really measuring and where the data is coming from. In addition to the information on the indicator design template, there are other aspects that will make it easier to communicate the performance information. Below I have summarized a number of best practice tips for performance reporting.

- Lead with commentary: Any report of performance data should focus on a short narrative comment or assessment of the performance that highlights what the data is telling us. I suggest that any performance report or communication should have a short headline summarizing the key findings in just one clear statement. In addition to the headline, it should have a three- to five-sentence natural-language explanation of the detail that also puts the results into the context of targets, benchmarks, key performance questions, etc. (see Fig. 9.3). The narrative commentary should provide an assessment

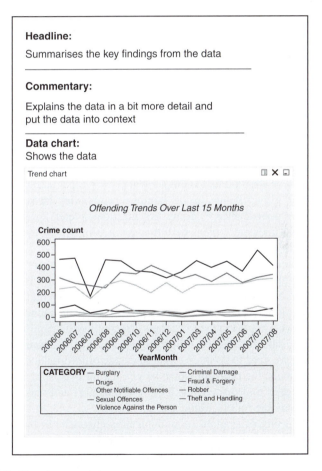

FIGURE 9.3 Narrative commentary.

on what this performance level means and whether there are any actions or initiatives being taken. Provided with this information, recipients of the data are now able to understand and interpret performance levels. The narrative comment is usually provided by the owner – the person(s) or function(s) responsible for the management of the strategic element that is being assessed. Leading with commentary engages people in the active review of indicators and provides a starting point for a discussion or dialogue about improvement.

- Visualize data in graphs: Performance data should be made easily understandable. In his book on data visualization, Stephen Few writes, 'Contrary to popular wisdom, the data cannot speak for themselves. Inattention to the design of quantitative communication results in huge hidden costs to most businesses. Time is wasted struggling to understand the meaning and significance of the numbers – time that could be spent doing something about them.'[24] One way to improve the communication of data is to use well-designed graphs and charts. Generally speaking, line graphs or bar charts seem to work well. They allow organizations to show past performance levels and allow inclusion of target lines and benchmark information (see Fig. 9.4). Many organizations now produce performance dashboards with traffic lights that provide at-a-glance assessment of the performance. Color coding and traffic lighting is very intuitive and useful for most people. However, beware that there are a lot of people who have difficulties distinguishing colors (especially the difference between red and green), which is better known as color blindness. It is estimated that about 8% of males and 1% of females have difficulties with color vision impairments, and therefore, it may be appropriate to complement or even replace color coding with symbols or icons (thumbs-up or thumbs-down, smiley face, etc.) in order to indicate performance.[25] Some organizations prefer speedometer-style displays that indicate current performance in comparison to the targets or expectations (see Fig. 9.5).

- Provide numerical data mainly in appendices: Use numbers and tables as supporting information (e.g. in appendices). The analogy I use here is a newspaper. Newspapers wouldn't just print tables of data on the front page. Instead, they start with a headline, a top-level comment and maybe a graph. Anybody who wants to read the details can continue on page 7 to find some underlying data and maybe follow a Web site link to more data. This is something we need to do more often. In an age of information overload, we shouldn't make the problem worse by circulating masses of raw data. The best practice is to provide some high-level numbers and make the remaining data available for electronic access.

- Use a good mix of available communication channels: In order to choose the right channel and format, it is recommended to consider the needs of the information participants – What questions do they want to have an answer to? This allows organizations to customize their reporting and

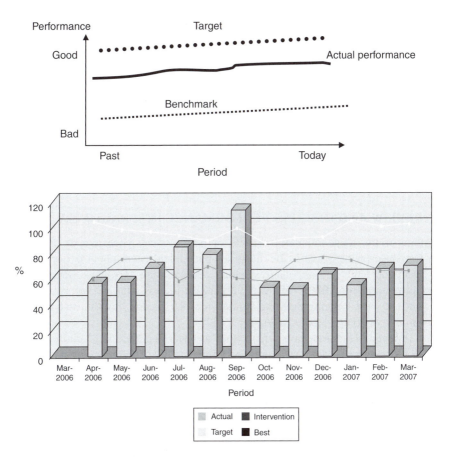

FIGURE 9.4 Line graph display.

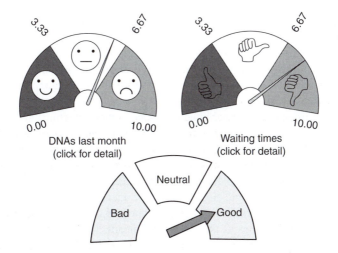

FIGURE 9.5 Speedometer display.

communication to the different audiences. Performance information can be communicated using various channels, including[26] different forms of print media (reports, newsletters and publications); advanced computer technology (e-mail, on-line Internet/intranet systems and video conferencing); highly visible means, such as the placement of wall charts with performance information in appropriate work areas; as well as interactive, group-oriented mechanisms (town hall meetings, performance meetings and focus groups), among others.

REVIEW AND DISCUSS PERFORMANCE INTERACTIVELY

One of the most important aspects of performance measurement and performance management is the dialogues that should occur about performance. Dialogue leads to joint understanding, insights and learning. IBM's Dean Spitzer argues, 'This is where the otherwise lifeless data and information is infused with meaning and transformed into knowledge, insight, and wisdom through ongoing, interactive learning.'[27]

Meetings that focus on performance and learning are a key building block of a performance-driven culture. Unfortunately, there are many organizations where no such meetings ever take place. It is surprising that I find it rare to see a group of people getting together to constructively discuss performance. If performance-related meetings take place at all, they tend to be dominated by lengthy and boring presentations of meaningless data. They are often called performance reviews. As the name 'performance *review* meeting' suggests, most of these are focused on past performance, often with a heavy bias toward financial indicators. They tend to be centered on budgets. One of the key questions is whether the budget was achieved in the last quarter.

Even worse, much time is wasted making excuses about why the performance targets weren't met, often shifting blame from one individual or department to the other. Little time is spent thinking about the future and how the performance drivers have to be managed to improve performance in the next quarter. Far too often, organizations seem to allow performance meetings to evolve haphazardly. Just because specific performance indicators are available, they are discussed and reviewed.

There is often a lack of focus and purpose in these meetings and they tend to be a mix of strategic and operational reviews. In those cases, the discussion often fluctuates between the highest level strategic issues to the minute details of operational and project-related issues. A lack of structure means that the meetings often go off course. Regularly, operational issues or 'firefighting' takes over and pushes the strategic discussion off the agenda. A lack of meeting discipline means that these meetings often start late and overrun and people tend to join or leave the meetings as they like. This means, when it comes to decision making (which is rare), some key decision makers might not be there to make them. Overall, people tend to see these meetings as a complete waste of time.

Below I outline three different scenarios of meetings used to discuss organizational performance (see also Fig. 9.6). Which one most resembles the process used in your organization?

- 'On trial' reviews: This resembles a court of law, whereby individuals are required to present their 'numbers' and explain to 'the boss' and other individuals present why some are good and, particularly, why some are bad. It is a tense atmosphere of 'trial by presentation'. If any executive cannot deliver a glossy PowerPoint presentation and, most importantly, satisfactorily answer penetrating questions about his or her department's performance, then he or she is likely to be humiliated and chastised by the boss, remanded in custody pending an appeal at the next meeting or added to the list for execution. The whole activity is similar to prosecutors and defendants arguing about who is to blame for the 'bad news'. I have been a witness to several meetings that typify 'on trial' reviews, and each time, I have

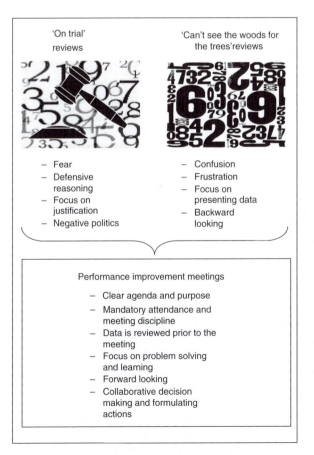

FIGURE 9.6 Toward better performance review meetings.

been appalled by the tension and dysfunctional behavior that this type of review invokes. Clearly, there is no spirit of cooperation among individuals simply because this is a struggle for survival – each person or department pitted against others. This way of conducting performance reviews is very closely linked to the aim of 'controlling people's behavior', discussed in the previous chapter. 'On trial' reviews destroy any performance culture that is conducive to collaboration and learning, and instead bring out an array of self-centered characteristics, negative politics and a focus on compliance.

- 'Can't see the wood for the trees' review: This is more like a random walk in the park – the discussion could go anywhere. Individuals present their 'numbers', but there is such a plethora of them that a somewhat-random debate then occurs about the causes of particular good and bad 'numbers', especially the potential causes of specific unusual 'spikes'. This results in these meetings tending to go into too much detail. The outcome of the debate tends to be to cut the discussion short (because the agenda is long and its planned timings have already overrun) and move on without making any strategic decisions because too much time has been expended on deliberating the minutiae. The issues that are being discussed are not put into the context of strategic intentions. Some strategic considerations might get discussed, but they seldom seem to get resolved with practical actions that are agreed. One manager compared this kind of meeting to driving a car where you have lots of data and information from the dashboard, but no idea where you are or where you are heading. This type of meeting is common in a 'reporting-only' culture, where the emphasis is on data collection as opposed to learning.

- 'Performance improvement' meeting: Participants know exactly what the agenda and purpose of the meeting will be and how the different elements being discussed fit into the strategic plan of the organization. Everyone who is supposed to be there is present and the meetings are run in a disciplined manner. Performance data (quantitative and qualitative) has been circulated in advance and individuals present and discuss only the issues resulting from the data. Participants focus on problem solving, decision making and formulating actions. Most importantly, the whole emphasis of the meeting is on dialogue and making *collective* decisions about what actions need to be taken to improve future performance. A performance improvement meeting is a sign for a performance-driven culture.

Obviously, there is territory in between these three extremes, but I am certain that many employees in many organizations will recognize some of the symptoms of the 'on trial' and 'can't see the wood for the trees' review meetings. If so, I believe they should reconsider the way they approach this vital process and, therefore, suggest that they need to have a debate among them about how they could move toward performance improvement meetings. For a start, the name 'review' meeting automatically focuses the attention on past performance. Whereas it is important to look at past trends and see how this might give us insights into future performance, many performance 'review' meetings look at only the past to review nonrelevant data or are too concerned about finding excuses and shifting blame, instead of concentrating on future performance and decision making.

CREATE DIFFERENT MEETINGS FOR DIFFERENT PURPOSES

In order to avoid the problems and confusions with performance-related meetings, I suggest creating four different and distinct types of meetings to discuss performance in an organization. Balanced scorecard pundits Bob Kaplan and Dave Norton agree that there is a need to split operational and strategic meetings.[28] Figure 9.7 shows the four meeting types I suggest establishing, namely, strategy revision meetings, strategic performance improvement meetings, operational performance improvement meetings and personal performance improvement meetings. These meetings are interdependent and the content and outputs influence each other. However, each of these meetings I propose here has its own clear purpose and each of them differs in terms of time horizon, frequency, outputs, focus and supporting performance information (see Table 9.1 for a summary). Let me briefly discuss each of these meetings in more detail.

- Strategy revision meetings: These are meetings that are used to revise and renew the strategy. Here the strategy and its underlying assumptions are questioned and newly shaped. In Chapters 1–4, I have discussed the different tools and techniques that can be used to design and revise your strategy.

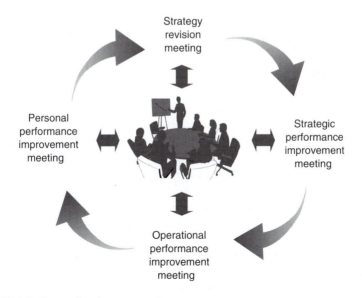

FIGURE 9.7 Proposed performance meetings.

TABLE 9.1 How the Performance Meetings Differ

	Strategy revision meeting	Strategic performance improvement meeting	Operational performance improvement meeting	Personal performance improvement meeting
Purpose	To revise and renew the strategy	To discuss the execution of the existing strategy	To discuss and respond to the operational issues	To discuss individual performance in the context of the organizational strategy
Time horizon	Long term (1–3 years)	Medium term (1–6 months)	Short term (1 week to 1 month)	Mid to long term (6–12 months)
Frequency	Annually	Monthly to quarterly	Weekly to daily	Annually to 6 monthly
Outputs	A new or revised value creation map and value creation narrative	A fine-tuned strategy execution plan, revised cross-functional objectives, revised operational activities and projects, revised budget allocation	Refined operational plans, solved short-term issues	Individual performance plans, individual development plans
Participants	Executive team, leader of the performance management team, performance analysts	Executive team and directors, departmental heads, performance management team members, performance analysts	Managers, departmental and functional supervisors and personnel	Everybody in the organization with their line manager
Focus/tools	Stakeholder analysis, PESTL analysis, scenario analysis, resource assessment (see Chapter 1–4)	Key performance questions, risk log, performance forecasts	Functional performance and issue management, project reviews, short-term budget reviews	Personal development log
Performance information	External and internal strategy analysis and reports	Key performance indicators (internal and external)	Operational performance information (internal)	Relevant strategic and operational performance indicators

In most government, public sector or not-for-profit organizations, these meetings would take place on an annual basis (it is rare that the environment is so dynamic that more frequent meetings to revise the strategy are required). The time horizon of these meetings is to look between 1 and 3 years ahead. The objective of these meetings is to agree on a new or revised value creation map and value creation narrative. The meeting is an opportunity for the executive team and the directors to get together and agree on their new or revised strategy. The executive team would take insights from various strategic analyses and performance data to firm up their strategy. It is usually recommended to also have the leader of the corporate performance management team and relevant performance management analysts in the meeting. These individuals can provide answers to any data queries and analyses. Strategy revision meetings tend to be held off-site and usually last between 1 and 2 days. As with all these meetings proposed here, the emphasis is not on data presentation, but on decision making and reaching strategic agreement.

- Strategic performance improvement meetings: These meetings have the purpose of discussing the execution of the existing strategy. Here, the overall strategic assumptions are not questioned; instead, the meetings take place to fine-tune elements of the strategy and to revise the strategy execution plans. Strategic performance improvement meetings are there to revise the operational activities of the strategic objectives on the value creation map. This would involve decisions about reallocating resources and refocusing projects. In these meetings, the value creation map and, in particular, the key performance questions guide the agenda, and performance indicators are used to guide the decision making. The time horizon of these discussions is medium term, meaning between 1 and 6 months ahead. Usually, these meetings would take place on a monthly basis and are attended by the executive team, together with directors and head of departments. Similarly to the strategy revision meetings, I would recommend that members of the corporate performance management team and relevant performance management analysts also attend the meeting to provide answers to any data queries and analyses. Strategic performance improvement meetings can also be used to model and test assumed causal relationships between different strategic objectives. From my experience, strategic performance improvement meetings are the type of meeting that is the rarest in organizations. At the same time, I believe that they are a core element of a performance-driven and strategically focused organization.

- Operational performance improvement meetings: These are meetings to discuss and respond to short-term operational issues. Often called 'performance clinics', they represent the frequent forums in which departmental managers and functional supervisors and personnel get together to talk about the 'burning issues'. In some organizations these meetings take place on a daily basis and in others on a weekly or twice-weekly basis. The discussions that take place and the decisions made in these meetings have a short time horizon (a week to a month). The meetings can focus on specific operational performance issues or can focus on project performance. Operational performance improvement meetings are the engine rooms of an organization in which the operational decisions are made and in which any short-term operational issues are discussed and resolved. A good example of an operational performance improvement meeting comes from the police departments in New York City (NYPD) and Philadelphia (PPD). So-called COMPSTAT meetings, which were initially developed in New York, take place to review the operational performance. The PPD describes the philosophy behind COMPSTAT as deceptively simple. It is based on four principles that have proven to be essential ingredients of an effective crime-fighting strategy:[29]

1. Accurate and timely intelligence: Effective crime-fighting requires accurate and timely intelligence. Officers at all levels of the police department must understand when (time of day, day of week, week of year) various types of crimes have been committed as well as how, where and by whom they have been committed.

2. Effective tactics: Having collated, analyzed and mapped this crime intelligence, the department's commanders must develop effective tactics for dealing with the problems it reveals. In order to bring about permanent change in crime conditions, these tactics must be comprehensive, flexible and adaptable to changing trends. They must also involve other law enforcement agencies such as the FBI, DEA and ATF; the prosecutors; the probation services; other city agencies not directly connected to law enforcement; as well as the public (community groups, Operation Town Watch, etc.).

3. Rapid deployment of personnel and resources: Once a tactical plan has been developed; the deployment of personnel and resources must be rapid and focused. To be effective, the response to a crime or quality-of-life problem demands that patrol and special units coordinate their resources and expertise and act with a sense of urgency.

4. Relentless follow-up and assessment: All action must be relentlessly followed up and assessed to ensure that the desired results have been achieved. This is the only way of ensuring that recurring or similar problems are dealt effectively in the future. The crime information is the basis for weekly meetings where the police commissioner and his or her entire top management team plan and coordinate the department's fight against crime. These meetings normally take place on Thursday mornings, begin at 7:30 a.m. and last for about 3 hours.[30]

- Personal performance improvement meetings: Most of the dreaded personal performance and development reviews that take place in organizations are purely administrative human resources (HR) tick-box exercises. The atmosphere is cringeworthy and the outcomes are not very constructive.

At most they tend to produce records of suggested training needs, which are sent to the HR department and only ever see the light of day again at the next round of meetings. Here, I suggest a different kind of meeting. Personal performance improvement meetings are the last missing element in a performance-driven culture. They should be forums in which employees and their line managers can discuss the strategic priorities for the next year. The time horizon for these meetings tends to be between 6 and 12 months and they usually take place on an annual basis. More recently, I have seen organizations that have successfully introduced 6-monthly personal performance improvement meetings. These meetings are a great opportunity to engage everybody in the organization in a strategic discussion and ensure any personal objectives, performance plans and development plans are aligned with the overall priorities of the organization.

While each of these four meetings has its own purpose and character as outlined above, there are some characteristics that all four of these meetings should share. Let me outline these characteristics in the following section.

CHARACTERISTICS OF PERFORMANCE MEETINGS IN A PERFORMANCE-DRIVEN CULTURE

Below I outline some common characteristics of performance meetings that would help organizations to get away from performance reviews that create dysfunctional behaviors and frustrations:

- Name the meeting appropriately: Take the word 'review' out of the name of the meetings. The main purpose of all four meetings is to improve future performance. Insights from the past can help us with decision making about the future, but it can't be the main focus of the meeting. Ensure that the name of the meeting reflects the purpose.
- Make the attendance mandatory for the key people: If the meetings are used to make the critical decisions about future performance, then it is important that all key decision makers attend. The police departments in New York and Philadelphia arrange their operational performance improvement meetings early in the morning, so participants are unlikely to have other commitments. Others organize their meetings a year in advance to make sure people get them into their diaries early enough without any conflicts.
- Maintain meeting discipline: Circulate the agenda in advance of the meeting, start and finish the meeting on time, follow the agenda, expect people to apologize if they can't attend, reach agreement on the action points and next steps and circulate minutes shortly after meetings.
- Create an atmosphere of purpose, trust and respect: The atmosphere in all these meetings should be purposeful but relaxed and friendly. Mutual trust,

respect and support lead to personal commitment, joint decision making and learning. Instead of a blame culture, the focus is on future performance, dialogue, decision making and actions. A chairman ensures that the agenda items are fully discussed and that any dialogue is constructive and aimed at improving future performance.

- Encourage dialogue: Dialogue is an enabler for learning. Through dialogue, individuals seek to inquire, share meanings, understand complex issues and uncover assumptions. In other words, dialogue is what facilitates evaluative inquiry learning processes of reflection, asking questions and identifying and clarifying values, beliefs, assumptions and knowledge.[31] Dialogue, as opposed to discussion, has the goal of understanding, not competition.[32] Dialogue requires the suspension of defensive reasoning and is about learning for change. It empowers individuals to share their thoughts and be heard, in order to reach joint conclusions. It is the responsibility not only of the meeting chairman but also of every participant to encourage dialogue.

- Never loose sight of strategy: The performance meetings should all link back to the strategy. All of the performance improvement meetings are opportunities to ensure everybody understands the strategic priories and the strategic map. I therefore believe that the value creation map should be used to guide the meeting. In the personal and strategic performance improvement meetings, the strategic map should provide the structure or agenda of the meetings. Many organizations now divide their strategic objectives and spend the meetings looking at a subset of their strategic objectives. If this is the case, then it still makes sense to briefly review the entire strategy map at the beginning of the meeting.[33] For the operational review meetings, I suggest to always start with the strategic map and from there drill down into the objectives and the operational issues. This ensures that people never loose sight of the strategic context of any operational issue.

- Use and value performance indicators: Performance indicators inform the decision making in these meetings. Participants take responsibility for analyzing the available performance data prior to the meeting with the aim of answering the posed question(s). The meetings are not there to present data but to tackle issues and decide on actions. If possible, any data-related issues should be resolved prior to the meetings. For that purpose, data analysts should work closely with leaders who are seeking answers to their performance questions.

- Ensure that collaborative decision making and learning takes place: It is important to capture everyone's opinion and to openly discuss the different points of view. This facilitates a better-informed debate and conflict resolution and enables collaborative decision making and mutual agreement on next steps and actions. In this kind of environment, it is acceptable to say 'I don't know the answer', instead of finding any answer for the sake of it. Decisions are made together and actions are agreed on, captured in the minutes and then followed up at the next meeting.

The characteristics outlined here provide the ingredients for successful performance meetings. It will, however, take time and efforts by everyone involved to make them work. Next I share a case study of an organization that was able to create a performance-driven culture as part of its strategic performance management initiative.

Case study: Creating a performance-driven culture[21]

While there are increasingly good examples of government, public sector and not-for-profit organizations that have successfully created a performance-driven culture, one of the best examples I have personally experienced comes from a commercial company. Here I want to share with you how Fujitsu Services has successfully created a performance-driven culture. Fujitsu Services is one of the leading IT services companies in Europe, the Middle East and Africa. It has an annual turnover of £1.74 billion, employs 14 500 people and operates in over 20 countries. It designs, builds and operates IT systems and services for customers in the financial services, telecommunications, retail, utilities and government markets. Its core strength is the delivery of IT infrastructure management and outsourcing across desktop, networking and data center environments, together with a full range of related services, from infrastructure consulting through to integration and deployment.

In Fujitsu Services, the helpdesks provide a critical function. These helpdesk call centers represent an integral part of service delivery and the primary point of contact for customers. If you are a customer who has outsourced your IT infrastructure management to Fujitsu Services, the helpdesk would be your point of contact if anything goes wrong or if you experience any problems with your computer hardware or software. Helpdesk agents can then either solve the problem or pass the work on, for example, to an engineer who then comes out and fixes the problem. It is often argued how call centers are changing the way companies communicate with customers and that they are a strategic asset in delivering exceptional service quality. Many organizations believe they are using their call centers to differentiate their product or service offering, to build and maintain customer relationships and to drive customer satisfaction.

The reality, however, is often very different. I am sure most of you can relate to the aggravation that is often caused when customers try to contact call centers or helpdesks. It often starts with a finger ballet to communicate with the interactive voice response (IVR) system and then endless queuing listening to the same irritating piece of music, and when we finally speak to someone they can sometimes be abrupt and unhelpful. Instead of treating call centers as service providers, they are often treated as unnecessary cost centers that have to be squeezed for efficiency. In many cases, this is due to outsourcing service level agreements, which specify performance targets of everything that is easy to measure, such as queuing time, the number of calls taken or average call duration.

In 1999, there was a growing realization at Fujitsu that the traditional approach to performance management was failing customers. Operating in the

IT outsourcing sector, Fujitsu found it almost impossible to differentiate itself in a very aggressive marketplace. A functional focus resulted in a lack of cohesion and fragmentation. Not dissimilar to other call centers, many client accounts were operating at contractual obligation levels and no higher, while 15% were at critical levels of dissatisfaction and were unlikely to be renewed. Furthermore, the turnover of frontline call center staff was 42%.

The message was stark for Fujitsu – it had to rethink its strategic performance management approach if it wanted to stand out from the crowd. What Fujitsu found was that the traditional way of measuring and managing performance stood in the way of a new strategic approach toward performance management. Fujitsu changed both the way that it approached performance measurement and performance management. In addition, Fujitsu saw this as an opportunity not only to redesign the organization but also to change the way Fujitsu worked with its customers. It was clear that customer satisfaction had to be a given. However, what Fujitsu wanted to change was its relationship with its customers – from service level contracts to a partnership model where customer success became a new goal. For this, it was critical to understand what was creating value for customers and what was not.

Fujitsu recognized that information about what was creating value for its customers had to come from its frontline agents since they are the ones speaking to customers all day long. However, the way performance was measured – with a strong focus on efficiency measures – prevented call center agents from spending time 'listening' to customers. All focus was on speed and numbers.

The first step Fujitsu took was to remove these measures from frontline employees to avoid the 'measurement trap' and prevent dysfunctional behavior. Call duration and the number of calls are still important indicators for managers to ensure the correct levels of resourcing, but they are wrong measures to influence the behavior of frontline agents. Fujitsu realized that if frontline agents are measured and rewarded on overall service delivery, they are the ones who can help to improve exactly this. They can provide critical information about service shortcomings, possible bottlenecks and future innovation. For that reason, Fujitsu changed its approach and started to treat call center agents as knowledge workers and began to leverage their knowledge for process and product innovation.

In order to create the context for knowledge work, the second step was to establish what I call an enabled learning environment. Fujitsu redesigned its management approach with a new emphasis on people, the problem-solving process and value creation. This involved a change in management style, with leadership principles based on intrinsic motivation and the creation of possibilities for others to succeed in a way that provides choice, not ultimatums. It involved the identification of training needs, the deployment of new skills and the reorganization of roles and responsibilities. The hierarchy within Fujitsu was essentially turned upside down. The role of managers was changed from one of authority to one of support. The central responsibility for them became the provision of the necessary knowledge and tools to allow frontline staff to handle the needs of the customer and assume responsibility for the end-to-end service, even if that service left the confines of the helpdesk at Fujitsu and was transferred to other client suppliers.

Today, dedicated frontline teams take on the role of establishing how they add value to their clients. They address questions such as 'What do our customers want to achieve?' and 'What is Fujitsu's role in this?' Its new strategic performance management approach enabled Fujitsu to move from a make-and-sell mentality toward a sense-and-respond mentality.[22] To understand how Fujitsu is creating value for their customers, frontline agents create a value creation map – a visual representation of the value proposition to their customers and the key competencies and performance drivers required from Fujitsu to help deliver the value proposition. In a so-called 'intervention process', frontline agents analyze the customer requirements and map out how they can help to deliver these. This often involves a visit to the customer sites to better understand their environment, working conditions and value proposition. Subsequently, the frontline agents design appropriate performance indicators that they own, review and act upon.

One of Fujitsu's customers is an airline company that has outsourced its IT management to Fujitsu. Airline employees would ring the helpdesk up if they experienced any problems with their IT equipment (e.g. printer doesn't work or servers are down). The success measures for the helpdesk team that handles the airline calls will be the overall service rating from the airline that has the IT infrastructure been managed satisfactorily by Fujitsu, instead of 'have the calls been handled within 2 min'. Frontline employees in Fujitsu now analyze and classify incoming calls in order to understand whether they are 'creating value' or 'restoring value'. The latter might be preventable by improving processes as part of Fujitsu's service delivery, for example, an engineer didn't turn up soon enough to fix an essential ticket printer at the airport and the customers are chasing up.

Frontline agents now look at what kind of calls they are getting and see what they tell them about their overall end-to-end service delivery. They might get calls because other parts of the business are not delivering, and therefore, customers are chasing their products. Trying to knock off a few seconds to optimize such calls would clearly be the wrong thing to do; instead, this information needs to be passed on in order to improve performance along the entire value chain. Cross-functional performance improvement meetings are used to explore how overall service delivery can be improved, and the input from frontline agents is of critical importance. Here new processes are established to ensure, for example, that the engineer turns up more quickly, the printers are replaced with more reliable printers or maybe clients are trained to fix essential equipment by themselves.

Sometimes, suboptimal processes in the customer organization are responsible for problems with the IT systems and are therefore preventable calls. In such cases, the information is fed back to the clients so that they can improve their own internal processes. In one case, Fujitsu discovered that many employees rang up to reset passwords at night, when no helpdesk was available for that client. This meant that they sometimes had to wait hours until the helpdesk agents were available again in the morning to reset a backlog of passwords. Instead of arranging 24-h helpdesk service, the solution was for the client company to change its processes and give some of its employees the ability to reset passwords when the helpdesk was not available. Under the old regime, there would have been no incentive for anyone in Fujitsu to suggest this approach.

For the airline company, helpdesk intelligence has managed to reduce queues at ticket offices, check-ins and boarding gates. Calls to the helpdesk have fallen by 30%, system availability has increased and client IT operating costs have decreased.

This new approach created completely new relationships between Fujitsu and its clients. Instead of operating only at the contractual obligation level according to efficiency measures specified in service level agreements, Fujitsu now operates at a partner level that allows mutual performance improvements. Commercial contracts between Fujitsu Services and its clients had to be restructured to realize mutual benefit from call reduction and mutual value maximization. The results of this change in the way performance is managed in Fujitsu Services are impressive. Today, Fujitsu achieves 20% higher customer satisfaction. It was further able to increase its employee satisfaction by 40%. Its staff attrition decreased from 42% to 8%, operating costs decreased by 20%, and contract renewal and service upgrades amounted to £200 million. Since implementation of its new strategic performance management approach, Fujitsu has won the National Business for the Best Customer Service Strategy and was awarded the European Call Centre of the Year award for the best people development program.

Today, Fujitsu is continuously redesigning its capabilities and offerings, not based on market intelligence but on customer knowledge and strategic performance data. Fujitsu recognized the potential of a new strategic management approach and applied it in a wider context. In addition to the helpdesk environment, these principles have now been applied to many other parts of the organization.

I believe that this case study demonstrates the power of an enabled learning environment and a performance-driven culture and how they can help to make strategic performance management a reality. It enables organizations to continuously learn and innovate, and therefore ensures long-term success. The time is right for more organizations to think about their strategic performance management processes and how to create an enabled learning environment.

SUMMARY

- In this chapter, I have outlined that creating the right organization culture is crucial in order to successfully manage performance in an organization. I have called this a *performance-driven culture*.
- *Organizational learning* is at the center of a performance-driven culture. I have introduced three levels of learning that need to be encouraged in a performance-driven culture:
  - *Single-loop learning*: Learning to improve existing strategy execution.
  - *Double-loop learning*: Learning to revise and challenge strategy and its assumptions.

- ○ *Triple-loop learning*: Learning to improve the performance information that is being collected.
- With organizational learning at the center, I outlined *four enablers of a performance-driven culture*: (1) a sense of community and a common purpose, which binds people together and provides the emotional commitment to success; (2) clear and accepted accountability for results and performance across the organization, which provides responsibly and ownership; (3) honesty and truth about performance results, which in turn create an atmosphere of trust and mutual respect; and (4) a clear definition of what a performance-driven culture is, which creates an understanding and acceptance of a performance-driven culture throughout the organization.
- I then introduced the *four key tools or building blocks* that will help to create a performance-driven culture, namely, (1) a strong performance-driven leadership, (2) a reward and recognition system that is linked to performance, (3) appropriate reporting and communication mechanisms and (4) interactive and constructive performance meetings.
- A strong *performance-driven leadership* can be created throughout the entire organization by (1) showing visible commitment to performance management, (2) explaining the role of performance management, (3) getting actively involved in performance management and (4) moving from inspector to supporter.
- An appropriate *reward and recognition system* ensures that performance has consequences. In order to align the rewards and recognition with performance, we need to celebrate success and don't 'hardwire' measures with compensation. In particular, we need to recognize and apply nonfinancial rewards. The most important one is to say 'thank you'.
- *Appropriate reporting and communication mechanisms* ensure that the messages about performance are received and understood. I provided a number of recommendations to improve our reporting and communication: (1) Lead with an explaining narrative commentary, (2) visualize data in graphs, (3) relegate raw data to appendices and (4) use a good mix of available communication channels.
- Creating *interactive and constructive performance review processes* engage people and generate a dialogue about performance that leads to learning, decision making and performance improvements. I proposed to create a performance meeting infrastructure consisting of four different meetings and produced guidelines of how to put them into practice.
- The four meeting types are *strategy revision meetings* used to revise and renew the strategy, *strategic performance improvement meetings* to discuss the execution of the existing strategy, *operational performance improvement meetings* to discuss and respond to short-term operational issues and *personal performance improvement meetings* to provide forums in which employees and their line managers can discuss the strategic priorities for the next year.

REFERENCES AND ENDNOTES

1. Spitzer, D. R. (2007). *Transforming Performance Measurement: Rethinking the Way We Measure and Drive Organizational Success*, p. 60; who talks about appropriate accuracy. American Management Association, New York.

2. Atkinson, P. (2004). Creating and Shaping a Performance Driven Culture. *Control*, 7, 21–26.

3. See for example: Dorgan, S. J., Dowdy, J. J., Van Reenen, J. and Rippin, T. M. (2005). *The Link between Management and Productivity*. London School of Economics, London; Risher, H. and Fay, C. (2007). *Managing for Better Performance: Enhancing Federal Performance Management Practices*. IBM Center for the Business of Government, US.

4. Risher, H. (2007). Fostering a Performance-Driven Culture in the Public Sector. *The Public Manager*, 36(3), 51–56.

5. See for example: Preskill, H. and Torres, R. T. (1999). *Evaluative Inquiry of Learning in Organizations*. Sage, Thousand Oaks, CA; Senge, P. M. (1990). *The Fifth Discipline: The Art and Practice of the Learning Organization*. Doubleday Currency, New York.

6. See for example: Argyris, C. (1999). *On Organizational Learning*, 2nd ed. Blackwell, Malden, MA.

7. See for example: Fried, A. and Marr, B. (2008). *Enacting Performance Measurement Systems in Strategic Learning Processes*. Conference paper, Strategic Management Society, Cologne; Argyris, C. (1978). Double Loop Learning in Organizations. *Harvard Business Review*, Sept–Oct, 115–25.

8. Ibid, Preskill and Torres (1999), p. 66 (see note 5 above).

9. Kaplan, R. S. and Norton, D. P. (1996). *The Balanced Scorecard – Translating Strategy into Action*, p. 17. Harvard Business School Press, Boston, MA.

10. See for example: Serving the American Public: Best Practices in Performance Measurement, 1997, National Performance Review, A1 Gore, Vice President of the United States, Washington. http://govinfo.library.unt.edu/npr/library/papers/benchmrk/nprbook.html

11. For a discussion about formative assessments, see for example: Black, P., Dylan, W., Harrison, C., Lee C. and Marshall, B. (2003). *Assessment for Learning: Putting It into Practice*. Open University Press, Maidenhead, UK; Black, P. and William, D (1998). Inside the Black Box: Raising Standards through Classroom Assessment. *Phi Delta Kappan*, London 139–144. Black and William recognize that standardized tests are very limited measures of learning and report that studies of formative assessment show an effect size (the ratio of the average improvement in test scores in the innovation to the range of scores of typical groups of pupils on the same tests) on standardized tests of between 0.4 and 0.7, larger than most known educational interventions.

12. See: www.fairtest.org. The Value of Formative Assessment. The Examiner, Winter 1999.

13. See for example: Mansell, W. (2007). *Education by Numbers – The Tyranny of Testing*. Politicos Publishing, London.

14. See for example UK school league table results at: www.dfes.gov.uk/performancetables

15. O'Neil, O. (2002). *A Question of Trust. The BBC Reith Lectures 2002*. Cambridge University Press, Cambridge.

16. Argyris, C. (1991). Teaching Smart People to Learn. *Harvard Business Review*, May/June, 99–109.

17. Source: http://www.weyerhaeuser.com/

18. Source: http://www.weyerhaeuser.com/ (August 2008)

19. Based on their 2007 employee climate survey.

20. See for example: Spitzer, D. R. (2007). *Transforming Performance Measurement: Rethinking the Way We Measure and Drive Organizational Success*. American Management Association, New York, p. 122; who talks about appropriate accuracy.

21. See for example: Serving the American Public: Best Practices in Performance Measurement, 1997, National Performance Review, Al Gore, Vice President of the United States, Washington.

22. The inspiration for this anecdote came from: Austin, R. D. (1996). *Measuring and Managing Performance in Organizations*. Dorset House Publishing, New York, p. 120.

23. Aberdeen Group. (2005). *Workforce Management: Creating a Performance Driven Culture*. Aberdeen Group, Boston, MA.

24. Source: Few, S. (2004). *Show Me the Numbers: Designing Tables and Graphs to Enlighten*. Analytics Press. Oakland, CA. Another good book on the topic is Few, S. (2006). *Information Dashboard Design: The Effective Visual Communication of Data*. O'Reilly, Sebastopol, CA.

25. For more information see Prevent Blindness America: http://www.preventblindness.org

26. See for example: Serving the American Public: Best Practices in Performance Measurement, 1997, National Performance Review, A1 Gore, Vice President of the United States, Washington.

27. Source: Spitzer, D. R. (2007). *Transforming Performance Measurement: Rethinking the Way We Measure and Drive Organizational Success*. American Management Association, New York, p. 105; who talks about appropriate accuracy.

28. In their book *The Execution Premium: Linking Strategy to Operations for Competitive Advantage*, 2008, Published by Harvard Business School Press, Robert Kaplan and David Norton propose to split performance related meetings into (1) operational reviews, (2) strategy reviews and (3) strategy testing and adapting.

29. Source: http://www.ppdonline.org/hq_compstat.php (August 2008)

30. For more information please see: http://www.ppdonline.org/hq_compstat.php

31. See for example: Preskill, H. and Torres, R. T. (1999). *Evaluative Inquiry of Learning in Organizations*. Sage, Thousand Oaks, CA, p. 53.

32. See for example: Bohm, D. (1996). *On Dialogue* (Nichol, L. ed.). Routledge, London.

33. See also: Kaplan, R. and Norton, D. (2008). *The Execution Premium: Linking Strategy to Operations for Competitive Advantage*. Harvard Business School Press, Boston, MA, p. 236.

Chapter 10

Leveraging Performance Management Software Applications

While performance management shouldn't be about software, I firmly believe that no organizationwide attempt to measure and manage performance can work without the appropriate support of specialized performance management software. Having the right software application in place and using it appropriately can make a massive difference and can enable users to learn and improve performance. Another important point to add here is that performance management software applications are not just something for the techies; they are an important success factor for managing performance that managers and indeed everybody in the organization need to be concerned with.

The questions I address in this chapter include:

- How can software help us leverage the power of performance management?
- How can software help to engage people in managing performance?
- How can applications support data analysis and extracting insights?
- How can solutions help to produce a single integrated view of performance?
- How can we distinguish between the different products?
- What credible vendors offer performance management software applications?
- How can you create your own list of requirements before purchasing an application?

As with any attempt at automation, there is always a real danger that it 'dumbs down' the real power of performance management and that organizations which rely too heavily on automation risk letting the software do the thinking. The other real danger is that the whole initiative now becomes an IT project, instead of an initiative to measure and manage what matters.

Performance management software applications are clearly not a magic pill that will sort out all the performance management problems. Indeed, I have seen countless government, public sector and not-for-profit organizations implement software only to find that, once the initial excitement had worn off, they are left with a costly IT system and a slow realization that performance management is not about technology but the people and their processes.

Performance management software application should facilitate and enable everything I have talked about in this book so far.

Before I discuss the benefits of, and selection criteria for, performance management software applications, I would like to give you another word of warning. Even though software applications can be a powerful enabler of the performance management process, this won't mean that these applications can do the work for you.[1] I have hopefully made it unmistakably clear that the key to a successful performance management approach is to design unique value creation logics and to develop indicators that help you answer your key performance question in order to support your strategic decisions. There is no shortcut to this. My advice would be to run as fast as you can if software salesmen are trying to tell you otherwise. There are no ready-made templates and no magic off-the-shelf strategic performance management frameworks. And there are no lists of generic key performance indicators that you can just plug in and play.

The performance management software application, if implemented correctly following the principles outlined in this book, will allow you to unleash the full power of strategic performance management. It will give all employees access to customized performance management data in their preferred formats, and it will enable collaborations, powerful analyses and data integration to provide a single integrated view of performance. Performance management software applications bring performance management to life and can make every single aspect of it easier.

I am often asked why we need performance management applications when we already have Excel. I believe that when it comes to performance management, spreadsheet applications do more harm than good. Spreadsheets were not designed for performance management and have many severe shortcomings when it comes to data analysis, communication and scalability. Let me use the next section to explain why spreadsheets are not suitable for performance management.

WHY SPREADSHEETS CAN'T DO THE JOB

Shockingly, a number of recent research studies, both in the commercial world and in the government sector, have found that a vast majority of organizations are still relying on spreadsheet applications, such as Microsoft Excel, as their main tool for performance management.[2] The same studies also reveal that those organizations using spreadsheets were not satisfied with their performance management capabilities, whereas users of specialized performance management software applications were most satisfied.[2] Many respondents who were using spreadsheets believed they were inappropriate tools for performance management because they are too cumbersome, labor-intensive and unreliable. And many organizations are currently looking for appropriate replacements. Let me summarize the major disadvantages of spreadsheet-based performance management solutions, and with it hopefully delete them as a viable performance management software option from your head:

- *No scalability*: Systems quickly reach the capacity which desktop spreadsheets cannot handle. Performance management spreadsheets can grow into big documents with color coding, macros, calculations etc. I have seen various spreadsheet-based applications becoming slow and prone to crashes. Often, there was just too much data and complexity in the spreadsheet, which were not designed for that purpose.
- *Time-consuming to update*: Spreadsheet-based solutions are usually manually fed and updated. In one organization that I was approached by, a group of business analysts spent about 1 week every quarter 'updating the spreadsheets'. This is not only slow but also leaves immense room for errors. A KPMG study found that over 90% of the existing spreadsheets contain mistakes![3]
- *No collaboration and communication support*: Information kept in individual spreadsheets is not designed for collaboration or communication. The spreadsheets are often scattered around on different machines, and it requires enormous discipline to work from the same spreadsheet.
- *Difficult analysis*: Analysis is complicated for the reason that data is stored in individual spreadsheets; it is difficult and time-consuming to bring them together for analysis across more than one data set.
- *Risk of losing the corporate memory*: It is rare that organizations keep a track record of changes in data over the years. What tends to happen is that spreadsheets get updated or new spreadsheets are created without keeping a record of historic data. This means that there is a real danger that the corporate memory gets deleted and with it the ability to learn from past performance.

Spreadsheet-based solutions are not really a workable option for any organization that is serious about performance management. For organizations that want to unleash the full potential of performance management, there is no alternative to installing purpose-built performance management software applications.

THE POWER OF PERFORMANCE MANAGEMENT SOFTWARE APPLICATIONS

Implementing organizationwide strategic performance management initiatives requires IT support. Paper and pencil, or simple spreadsheet tools might be sufficient at the beginning when organizations start designing their performance management approaches. However, in order to make strategic performance management an integral part of the organization, automation will be necessary. The so-called performance management solutions help to integrate data from disparate sources, enable organizations to analyze the data across all

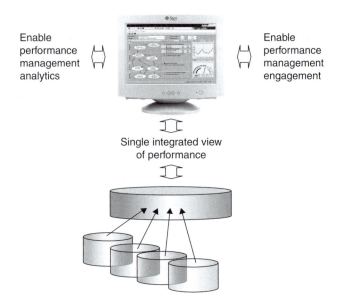

Enable
performance
management
analytics

Enable
performance
management
engagement

Single integrated view
of performance

FIGURE 10.1   Benefits of performance management software applications.

strategic elements and, most importantly, actively engage people in performance management through collaboration and communication features.

According to Professor André de Waal, one of the seven performance management challenges organizations need to address is to embrace information transparency in order to have the right information available at the right time, to make the best decisions and to take actions.[4] Overall, performance management software applications help to create a single integrated view of performance, they engage people in performance, and they allow everyone to analyze performance and gain insights (see Fig. 10.1). Below, I discuss each of the three areas in more detail.[5]

However, before I move on to discuss each of the key capabilities of software application in more detail, it is important to note one other thing: Software will always only be an enabler and enhancer. This means that if the foundations of the performance management approach are weak, the automation will be weak too.

Creating a Single Integrated View of Performance

The first huge benefit is that performance management software applications enable the integration and management of the performance data in an organization. Performance-related data is usually held in many different formats, in diverse databases and in different formats. Bringing the relevant data together in one data repository is the first important step in an automation process.

However, managers often believe that it is just a matter of connecting to the existing databases and then pulling out the data into a single data repository that underpins the performance management application. The problem is that the efforts necessary to integrate and collect data from disparate sources are often underestimated. And more importantly, a lot of the performance data required is not readily available in existing databases. From my experience, on average only about 20–30% of the required data is held in existing databases.

The first step, therefore, is to find out which information is relevant and required, whether the data already exists and, if so, where it is stored and how the data can be accessed. Most organizations have made significant investments in data warehouses, data marts and enterprise resource planning (ERP) systems, which mean that a portion of the required information can come directly from these back-office systems.

However, a significant amount of information will usually come from office applications, such as Microsoft Excel, where the data is stored in spreadsheets. There will also be an increasing amount of data from third-party providers that has to be fed into an application. Third parties could provide, for example, any benchmark information, customer satisfaction information, or brand and reputation data. Many organizations also outsource their employee satisfaction surveys. A large part of the organizational performance data may have to be entered manually into the system, either because it is nonexistent or because it is not stored in the available IT systems. Tapping into different data sources and creating automated feeds is not a trivial task and it is important, therefore, to ask yourself whether it is really necessary (and economical) to connect databases.

For performance management, the relevant data is usually pulled together into a single database or data warehouse, which creates a single integrated view of 'the truth'.

Using Performance Management Software to Engage People

One of the key benefits of an automated solution is that it enables organizations to engage people in performance management. Performance management can be brought to life through powerful communication and collaboration features.

One of the most important capabilities is the visualization of interactive value creation maps. These allow employees to view the strategy and understand not only the strategic logic but also the performance levels for the different strategic elements and objectives. Figures 10.2 and 10.3 show the value creation maps of organizations discussed in Chapter 4. Figure 10.2 shows an illustrative example of the value creation map of the Motor Neuron Disease Association with active performance gauges and Fig. 10.3 shows Belfast City Council's value creation map with color-coded fields. The elements of the value creation map are active and through a Web-browser interface display the latest performance assessments. Users can then click on these elements and 'drill down'

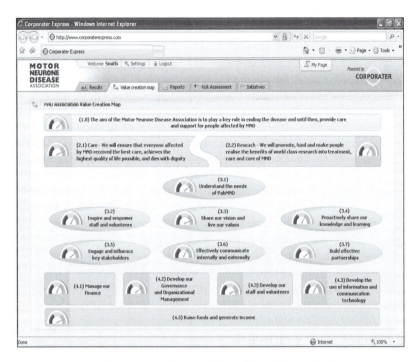

FIGURE 10.2   Interactive value creation map display with speedometer displays.

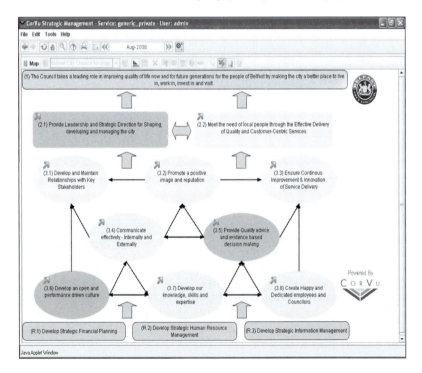

FIGURE 10.3   Interactive value creation map display with color-coded fields.

to the underlying performance indicators, view descriptions, definitions and comments as needed, and go on to analyze the data further. Figure 10.4 provides another example of a value creation map, visualized in a hierarchical tree diagram. Users can, for example, click on the bottom elements to extend the tree further or they can select different strategic elements to display information such as definitions, related action plans and indicator data on the right-hand side of the screen. Views can be created to suit any requirements and any format.

Performance dashboards create visually rich displays so that users can understand performance information at a glance. These displays are usually automatically updated based on the latest data available. Color coding and intuitive performance displays allow users to understand performance levels. Figure 10.5 shows a dashboard with color-coded tabs from the US Department of Labor and Fig. 10.6 shows an example of a police dashboard providing viewers with a top-level snapshot of performance. It is easy to turn these dashboards into performance portals with links to other documents and applications. From here, further 'drill-downs' are possible to select an indicator view or a view of initiatives that are linked to top-level dashboards.

Most of the applications available today are fully Web-based and provide Web-based access. This allows access to the latest performance data from anywhere in the world where you have access to an Internet browser. Security features such as usernames and passwords allow users to be identified. This also

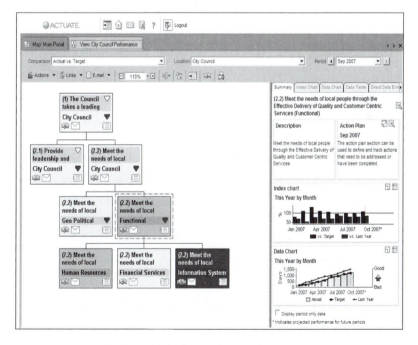

FIGURE 10.4 Interactive hierarchical value creation map display.

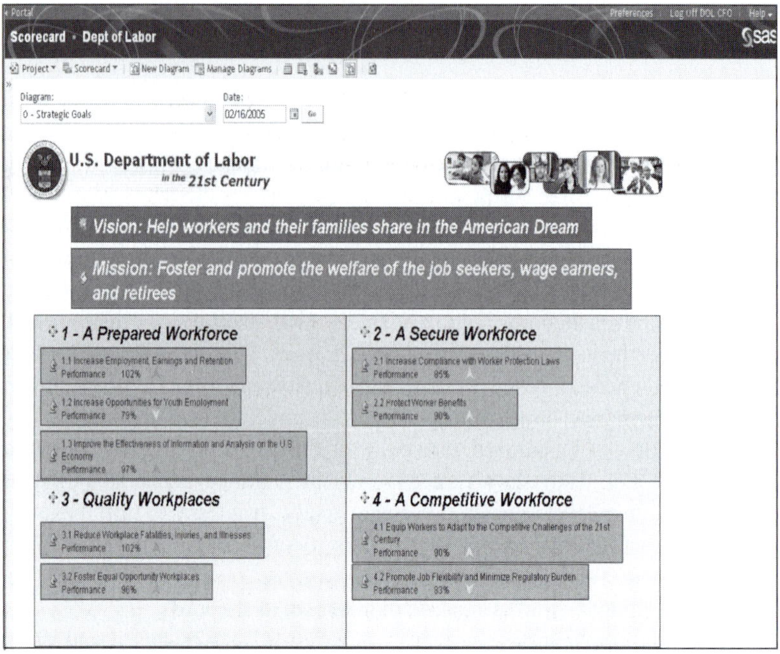

FIGURE 10.5 Dashboard (US Department of Labor).

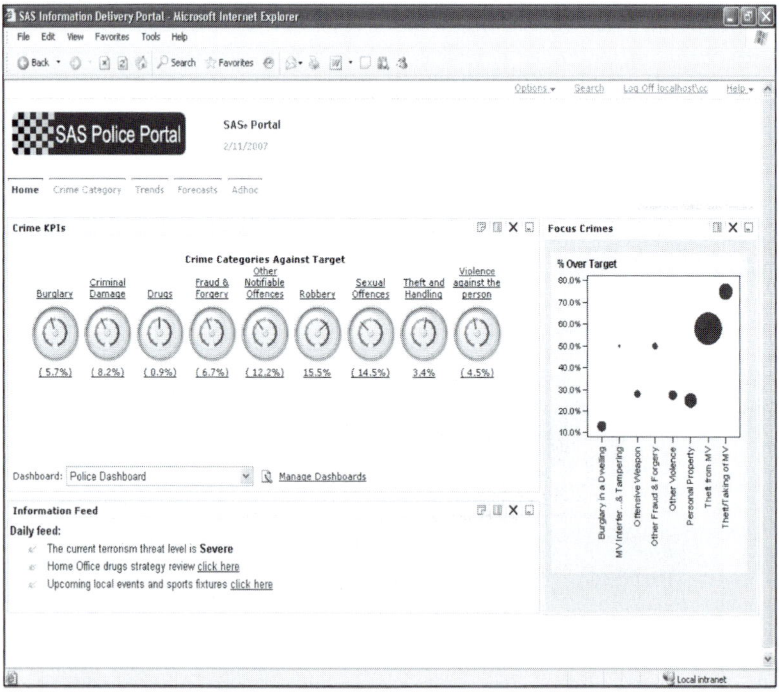

FIGURE 10.6 Police dashboard.

gives companies the option of providing subsets of the entire performance data only to selected groups in the organization. It would, for example, be possible to ensure that only branch managers in New York can view their branch data, and that the branch manager from London has no access to that data.

Once users have logged on, the system recognizes them and provides them with customized homepages that display the information relevant to the particular user. Instead of asking users to find the relevant information, it is easy to provide them with everything they need. In many cases, users are able to define the content of their dashboards. Figure 10.7 shows a personalized homepage dashboard with personalized performance information, links and alerts.

Most software applications make it easy to create different standard reports in which the data gets updated automatically. These can take any form or shape and can be freely customized for specific users or group of users. Senior executives might get a quarterly performance summary and operations managers might get weekly updates of a set of key operations measures in a trend view. It is therefore possible to create standard reporting templates, automatically generate the reports with the latest data and e-mail them out at given times to the selected participants. Figure 10.8 shows an example of a customized report in a briefing book format.

Performance management systems are able to provide automatically triggered exception alerts. If, for example, a specific measure reaches a predefined

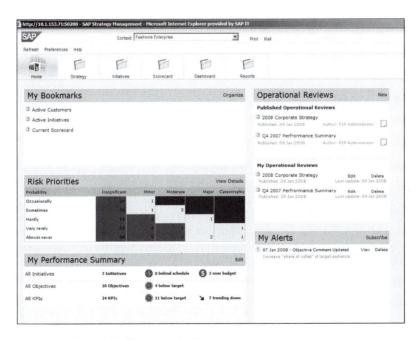

FIGURE 10.7   Personalized homepage dashboard.

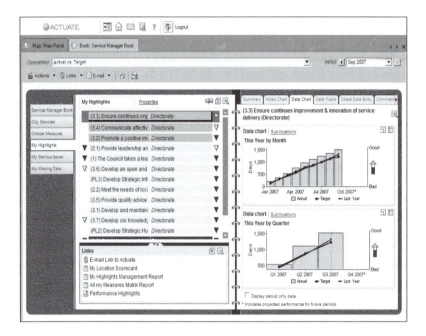

FIGURE 10.8 Customized report.

threshold, automated e-mails or SMS messages could be sent to individuals or to a group of people. Performance management software applications not only provide the functionality to push data to users, but also enable them to provide feedback and comments, or even start discussion threads around specific topics.

These applications can also remind people to update their performance indicators and even chase them up! Often they provide a simple Web interface in which people can update their performance indicators. Many of the systems also include in-built survey functionality, which allows organizations to automate the data collection for their staff or customer surveys. In addition, workflow capabilities can support initiatives and actions and provide a fast, automated (or ad hoc) way of collaboration and engagement.

In Chapter 5, I have discussed the importance of aligning your strategy and performance management with other activities such as planning, project management and risk management. Many of the available applications now provide fully integrated solutions which allow you to manage these projects, risks, budgets etc. as part of their application. See, for example, Fig. 10.9 which shows a project management module of a performance management suite and Fig. 10.10 which depicts a risk management module.

All these features make sure that performance management becomes real for people and help to engage people in managing performance, while taking much of the pain out of the process.

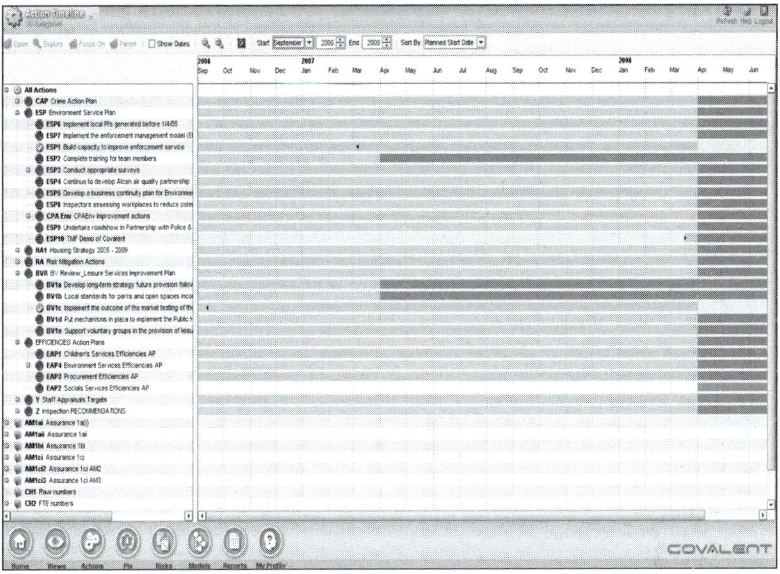

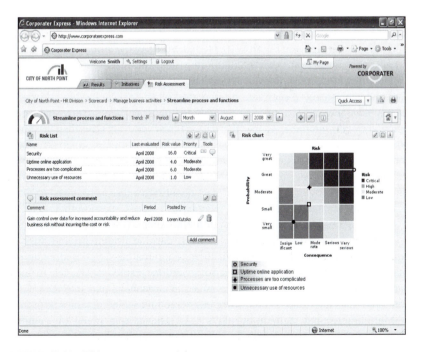

FIGURE 10.9   Project management module.

FIGURE 10.10   Risk management module.

Enabling Performance Management Analytics

The final major benefit of strategic performance management software applications is the ability to analyze performance data much more effectively and comprehensively. Providing everyone in the organization with the ability to apply some analytics to performance data is a powerful and key component of a performance-driven culture. The interactive drill-down capabilities described above are a simple and intuitive way of exploring and analyzing performance data. However, many of the performance management software applications provide much more sophisticated analysis and business intelligence (BI) features, such as the following:

- Sophisticated graphing
- Trend analysis
- Root-cause and impact analysis
- Modeling
- Correlations and regression analysis
- Multidimensional OLAP analysis
- Forecasting
- Simulation and scenario features.

Visualizing data in more graphical formats can be very powerful. In the previous section I have described how entire interactive value creation maps can be visualized. Nowadays, many of the software applications come with powerful graphic capabilities that go far beyond what ordinary spreadsheet applications can deliver. Figures 10.11 and 10.12 show, for example, color-coded geographical maps together with other analysis charts and graphs.

The other important analysis is impact or root-cause analysis or the modeling and assessment of correlations or regressions in causal models. If organizations have created causal value creation maps or other cause-and-effect logics, they are then able to use the data to 'test' and validate their assumed relationships. Figure 10.13 shows an example of a root-cause-analysis screen to test the impacts and causal relationships in models. In some of the more analytical applications, users are able to create simulations based on their cause-and-effect logics. However, a lot of quantitative data is required to make such simulations meaningful, and in most cases I would question their value.

Often, data has to be viewed from different perspectives and a sophisticated technique is needed to explore accumulated data. Multidimensional analysis tools usually perform this task. With them, data can be stored and examined in a multidimensional format similar to an ordinary spreadsheet, but in more than two dimensions.[6] These tools are linked to a graphical user interface (GUI) which provides the results on the computer screen presented in tables or graphs.

Multidimensional technology plays a significant role in BI by enabling users to make business decisions by creating data models that reflect the complexities found in real-life structures and relationships. It consolidates

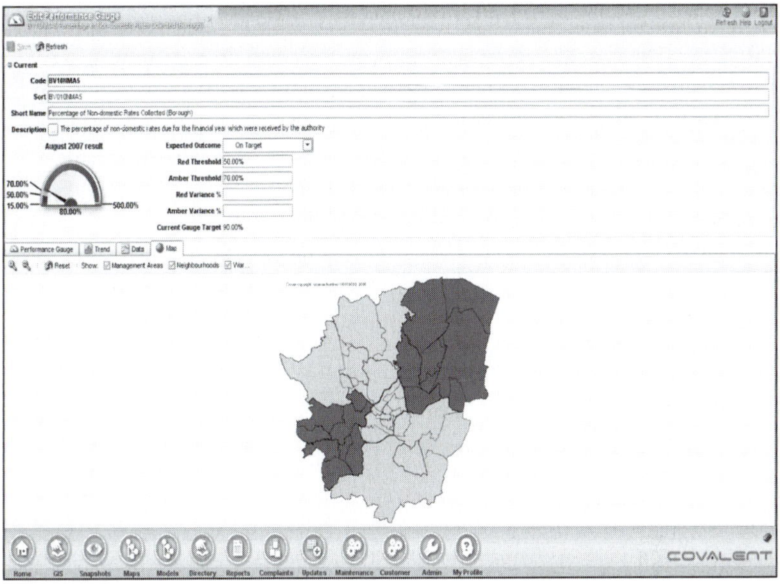

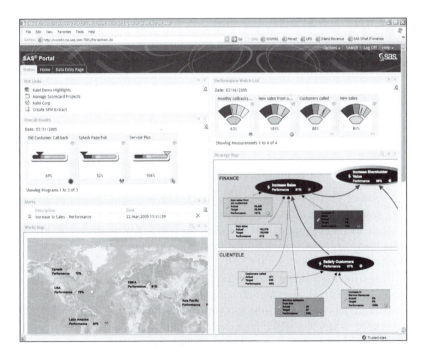

FIGURE 10.11 Color-coded Map.

FIGURE 10.12 Geographic and analytics data.

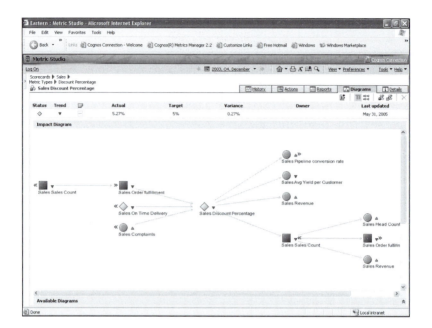

FIGURE 10.13 Screenshot: impact analysis.

and presents summarized corporate information from a multitude of sources. Multidimensionality, as a concept, can seem highly abstract at first, but it simply expresses the way we naturally think. The best way to grasp the advantage of multidimensional viewing is to think of a three-dimensional cube. To understand the benefit of the concept, I would like to provide the following practical example taken from an SAS Institute white paper (see also Fig. 10.14).[7]

The users might be interested in the sales performance of the organization's products. The three dimensions (i.e. product sales, region and time) might all be of interest to a number of users, but each might want to view the data from a different perspective according to each user's function. Several examples follow:

- A product manager might be interested in the performance of one particular product line in all regions over time (the 'product line perspective').
- A financial analyst might need to view the total sales results of all products in all regions within a particular timeframe, such as a calendar month (the 'financial perspective').
- A local manager might want information on sales results within a specific geographic region (the 'regional perspective').
- Finally, a market analyst might be interested in focusing on a single cell in the cube, a cell being the intersection of all dimensions at one point (the 'comparative perspective'). Typically, such an inquiry is undertaken for comparison purposes.

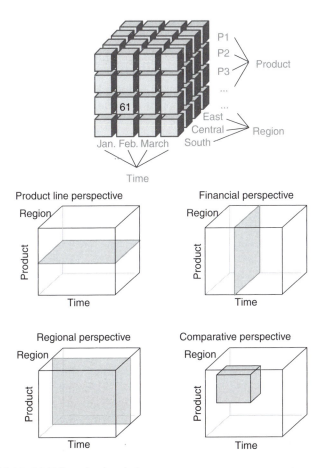

FIGURE 10.14 Multidimensional analysis.

These examples look at the same data, but each has a unique perspective. Cutting through the multidimensional cube to reveal various perspectives is often called 'slicing and dicing'.

Overall, performance management software solutions are not just big databases full of numbers. They also hold, for example, visualizations, definitions, descriptions, comments, discussion threads and action plans. These provide the rich contextual information that allows us to make sense of the data and turn it into actionable knowledge and learning, which is the key objective of strategic performance management.

SELECTING APPROPRIATE SOFTWARE SOLUTIONS[8]

The market for software solutions is growing rapidly and more vendors are trying to get their share of the multibillion-dollar analytic application market.

You just need to turn to the Internet and search on strategic performance management software solutions, and you will easily find in excess of 30 different vendors all of whom are willing to offer you a performance management application. Each vendor will claim unique advantages and features of their particular product, and each vendor will be able to provide credentials from satisfied users. Managers looking for an appropriate solution for their organization have often little to base their decision on or few tools to distinguish the various vendors. Below, I address these issues and provide you with a framework of how to select the most suitable solution for your organization.

Buy and Do Not Make

One of the first questions you will be faced with is: 'Do I buy a packaged application from a specialized vendor or do I build my own?' Some organizations choose to develop their own software. The 'create-your-own' solution allows organizations to address their unique needs and objectives; however, it is generally much more cost-effective to purchase a packaged application rather than to create your own.

Besides the cost factors, packaged applications are usually quicker to implement with vendors offering a wide variety of ancillary services such as conversion assistance, implementation training and system integration. These packages also tend to represent the cumulative efforts of many developers and customer organizations over a longer period of time, which usually results in better, more user-friendly applications than most attempts to create 'home-grown' applications. DaimlerCrysler, for example, initially decided to develop its own solution. The problem was that it took 5–10 people over 18 months to develop the first, dubbed application.

Reasons for choosing a 'home-grown' solution might be that functional criteria are not met by packaged applications or that there may be a BI software available within your organization that supports much of the functionality required. However, today's packaged applications are so good that I have never seen a valid justification for developing a home-grown performance management system. Unless you have a lot of unused IT resources, an extremely good understanding of strategic performance management functionality, and a lot of spare cash to burn, I would always recommend you to go for a packaged application.

Performance Management Automation Choices

Before you go on to automate your performance management actives, there are some bigger choices to be made about the following:

- Implementation: What is your desired scope of the software implementation?
- Integration: How deeply do you want to integrate the software with operational and transactional systems?

- Customization and IT support/skills: To what extent do you need or want the software to be customized?

Let me briefly discuss each of them in more detail:

- *Scope of the Implementation*: The scope of the implementation can be very different and various vendors have different views on how big (or little) the scope of implementation should be. The smallest implementation scope would be an off-the-shelf point solution that runs as a stand-alone application on a PC and might replace an Excel spreadsheet by offering more specific functionality. These packages are usually quite inexpensive and can be bought in ready-to-use form. The advantage is that you can start using the software straight away, usually by inputting the data manually.

  To increase the scope, vendors offer a broader set of applications which are likely to access a similar set of data. Those applications could, for example, include risk management, activity-based costing (ABC) or activity-based management (ABM) solutions, planning and budgeting application, project management applications as well as solutions for stakeholder-relationship management (SRM), process mapping, workflow management, customer relationship management (CRM) or business consolidation. The implementation scope broadens when organizations also try to support some of the other applications or solutions. The broadest scope would be an integrated strategic performance management suite (sometimes also called enterprise performance management or corporate performance management suites) that offers a wide range of the applications and functionality mentioned above. With an increasing implementation scope, there is also a rising need for technical solutions like data warehouses or multidimensional databases that hold the data for the applications.

  Organizations that are planning to purchase a software solution should consider the implementation scope, not only for the initial implementation but also taking into account future plans. You might want to start with simple reporting and analysis functionality but later expand the solution to other additional functionality. Many organizations already have applications for planning and budgeting, risk management, project management or CRM in place. In this case, they might want to ensure that the new application they are purchasing is compatible with the applications already in place. Quite a few of the stand-alone solution providers have entered partnerships with other third-party vendors in order to be able to offer native links to products like ABM or CRM software.

- *Degree of integration*: The degree of integration refers to how deeply the performance management software is integrated with underlying operational systems like service management systems, order entry, account systems, general ledger, purchasing, warehousing, human resource systems or general ERP systems.

  Some of the data required for the performance management applications usually resides in the underlying operational systems and databases. Full integration or 'closed-loop' integration would mean that the application is seamlessly integrated with the underlying operational systems and data can be fed automatically in either direction – from the underling operational systems to the application and vice versa.

The products on offer vary in their integration ability. Basic stand-alone applications might offer no integration capabilities with underlying systems and are not designed for automated data feeds but manual data transfer. These systems, however, might offer integration on an analytical level. This would mean users can analyze the existing data in the system through drill-downs, for example, from an aggregate high-level performance indicator to underlying indicators by querying a database. If the software application does not offer this capability, it can be achieved by using BI applications. Some performance management applications offer integration with BI tools. Other applications are designed for full integration and do not even allow manual data input. As discussed before, I think manual data input is always required and you might not need a lot of integration. It is important to really think about the level of integration you require.

- *Required customization and IT support*: You also need to look at the required customization and the required IT support. Not all products in the market place were developed as strategic performance management applications; some are more generic BI tools that allow organizations to customize their applications and more or less create their own applications. Some of the more advanced products offer things like wizards that guide users through parts of the design process for value creation maps or indicators, while other products offer no or little of such guidance.

There are several issues you might want to consider before making the decision. The advantage of a ready-made application is that it includes all the methodological intelligence as well as the expertise of experts. The downside might be that the methodology does not completely fit with the methodology your organization would like to implement. If your organization has specific requirements which cannot be found in any of the more advanced solutions, it might be a reason for using more generic BI tools. However, nowadays, most of the packaged applications offer enormous flexibility to be customized to your requirements.

Besides using more generic BI applications to create customized solutions, you can also use vendors or consultants to customize solutions for you. There are few management consultancies that specialize in customizing performance management applications by using their own or various third-party software components. This might be an option if an organization lacks internal expertise in methodology or has not got the internal IT resources to support the internal developments.

Having thought about all the above, you are still left with too many choices. So how do you know which of the different vendors to go with? How do you shortlist the number of vendors from over 30 to 3? What is the process of making the right decision about selecting the appropriate solution? In the next section, I present a framework to assist you in the decision process.

Performance Management Software Selection Framework

Selecting the right performance management software solution is a major decision for most organizations. The prices for software solutions vary enormously from a few thousand dollars to far over a million dollars. A realistic starting price lies at about $50 000, with reasonably sized organizations typically spending about $200 000. Making the wrong decision and buying the wrong software can not only result in a significant waste of time, energy and money, but can also undermine the entire strategic performance management effort and the credibility of the performance management approach you are trying to put in place.

The starting point for any selection process has to be the recognition that each organization has a unique set of requirements for their strategic performance management approach. It is therefore not possible to provide a list of requirements that is appropriate for every organization. Organizations differ in terms of size, IT infrastructure, communication style, required level of security, cash position, strategic design, IT literacy, in-house capabilities etc. All these have to be taken into account while determining the selection criteria of an appropriate software solution. For the purpose of developing a selection framework, this means that I can present the criteria you need to consider, but then you have to decide what you need and weigh each of the criteria to reflect your unique set of requirements.

Following the same logic, each of the different software solutions available has different strengths and weaknesses. The natural starting point is identifying what you really want and need and then see whether that is available from packaged applications. The easiest way to do this is to create a two-directional matrix in which you put weightings against each criteria; this matrix can then be used to compare the available software products against the organizational requirements (Fig. 10.15). In the following, I will define and explain each of the 11 selection criteria you should think about before choosing a performance management software application:

1. *Costs and pricing*: First, it is a good idea to check basic company information about the vendor as well as information about the software product. The main aspect here is the pricing, since prices as well as pricing models vary significantly. Here it is important to check not just the license fees but also the maintenance fees, which can fluctuate between 5% and 30% of the license fees. Software pricing is a complex issue and different pricing models might suit one organization better than others, for example, pricing per user versus pricing per package. However, software companies are often flexible in their pricing, and pricing models are subject to

Criteria	Weight	Product A	Product B	Product

Costs/Pricing
   Sub-criteria I
   Sub-criteria II
   ...
	0–10
	0–10			

Company/product
   Sub-criteria I
   Sub-criteria II
   ...
	0–10
	0–10			

Scalability needs
   Sub-criteria I
   Sub-criteria II
   ...
	0–10
	0–10			

User interface
   Sub-criteria I
   Sub-criteria II
   ...
	0–10
	0–10			

Flexibility needs
   Sub-criteria I
   Sub-criteria II
   ...
	0–10
	0–10			

Communication and
collaboration needs
   Sub-criteria I
   Sub-criteria II
   ...
	0–10
	0–10			

Security and access
control
   Sub-criteria I
   Sub-criteria II
   ...
	0–10
	0–10			

Technical Needs
   Sub-criteria I
   Sub-criteria II
   ...
	0–10
	0–10			

Etc.	0–10 0–10			
Etc.	0–10 0–10			
Score:				

FIGURE 10.15   Software-selection matrix.

negotiation. It is also important to consider training and implementation costs as they can drastically increase the overall price of solutions, but often remain initially hidden.

2. *Vendor information*: It might be good to understand the background of the company and the product as well as how many people work on the performance management solution. Very large software companies might have only few people working on their performance management application, which might be treated as a by-product. On the other hand, a small company that specializes in strategic performance management software might have more expertise and a larger client list. The size and global presence of a software vendor might be important if organizations plan to implement the application globally or across countries. Organizations might want to check the economic viability of the vendor considering recent collapses and mergers in this market.

3. *Scalability*: In order to assess the required scalability, it is important to consider the final implementation scope. An organization might initially automate only one department or business unit, but later plan to roll it out throughout the organization. There are three aspects of scalability: (1) The application should be scalable in terms of programming. It should, for example, be easy to add new cascaded value creation maps. (2) The underlying database should be scalable since the amount of data and information accumulates quickly. (3) The communication approach should be scalable so that it is easy to disseminate the information through the Web browser. Language can also be an issue for international organizations and they might want to check whether the application comes in various languages. I always recommend visiting some reference customers with similar implementation scope to get some actual feedback.

4. *User interface/data presentation*: Here you have to decide about your visualization and data presentation needs. Applications vary between very graphical to more text- and tables-based. One of the most important aspects is the display of value creation maps and cause-and-effect relationships. I recommend going for interactive and dynamic visualizations, where the underlying data is linked to the different elements and where the causal connections mean something. Some tools just display graphics without any underpinning intelligence, drill-down or impact analysis functionality. Dynamic maps allow you to use them as a powerful communication tool with traffic lighting and even provide the opportunity to mathematically test assumed relationships.

5. *Analysis functionality*: Tools offer different levels of analysis capabilities, stretching from simple drill-down capabilities to multidimensional analysis, complex statistical functionality, forecasting and even simulations. Organizations which require more complex analysis functionality often have tools for this already in place and have to decide whether to integrate or replace those. Analysis functionality also includes the number of

graphical displays (from bar charts to advanced 3D charts). Requirements in terms of charts and graphs depend on the indicators the organization tracks and their visualization requirements. For this discussion, it is especially important to include the business analysts in the discussion.

6. *Communication and collaboration*: The communication aspect of any strategic performance management implementation is very important. Organizations have to address issues such as: Do you want the software to be Web-enabled? Do you want users to be able to comment on any aspect of the strategy or do you want to restrict the commentary to a group, for example, managers responsible for certain aspects in the strategy. For the majority of implementations, it is important that the application integrates with the existing e-mail system so that alerts, reminders, assessments and comments can be sent to specific users. Most software solutions are able to trigger automated alerts, e-mails or SMS messages, which can be sent to individuals or groups indicating that certain areas of the business are underperforming and hence require action. Most applications allow you to assign owners (and persons responsible for data entry) in order to automate the data collection and remind them if data, comments or assessments have not been entered. You might want the software to support action and include activity or project management functionality that allows you to track progress against strategic objectives. Some organizations love and fully embrace this data-push concept and workflows, whereas others feel that such an approach is too intrusive and doesn't fit with their current culture.

7. *Flexibility and customization*: This is an important aspect and nowadays organizations are less willing to invest into applications that are not, for example, able to integrate with other applications. Many tools provide interfaces with reporting packages, ABC solutions, CRM or planning tools. Flexibility should also be provided in terms of methodology support. Many organizations have multiple performance measurement and reporting needs; besides, their strategic value creation map might also want to use the software for frameworks such as other business scorecards or assessment frameworks (e.g. Baldrige National Quality Award, EFQM Award, Deming Prize, Charter Mark, Investors in People etc.). It usually makes sense to use the same application for all the performance measurement and reporting needs.

8. *Security and access control*: You need to decide about the level of security needed in the system; some organizations are very open and happily share all aspects of their strategic performance with all employees, others require very tight security for some aspects of performance.

9. *Technical specifications*: The technical requirements depend on the existing information and communication infrastructure in your organization. Any new piece of software should support the existing desktop or network

operating system. For a strategic performance management application, it can be important to be able to extract data from existing data sources. This can be a major obstacle for any implementation. It is a good idea to involve the IT department in the discussion about technical requirements.

10. *Service*: Vendors offer different levels of service. Some offer no implementation support and, instead, partner with implementation consultants to provide this. Other vendors offer comprehensive services including their own implementation service, consulting, international service hot-lines etc. Organizations need to be clear how much support they want and whether the vendor or their partners can deliver this.

11. *Future vision*: Here the future developments and release frequency of the product are addressed, which might indicate the vendor's attention and commitment to the product. It is also important to understand the future vision of the software vendor, which will influence the direction of any future product development. In an ideal case, the future view of performance management would be similar for the vendor and your organization in order to ensure future compatibility.

Overall, it is important to involve different people in the process of developing the requirements for your strategic performance management solution. Organizations often fail to involve all key functions and end up with a solution that matches only half of their organizational requirements. When only IT people are involved, they typically look for the IT-specific capabilities and compatibility with the existing IT infrastructure; finance people usually look for financial capabilities and economically most sensible solutions; business analysts may look for the most comprehensive analyzing capabilities; and general managers may look for a good user interface and ease of use. In order to address all requirements, it is therefore important to involve members from all four groups in the decision process. My experience has taught me that the selection process is best led by members of the management team in close collaboration with business analysts and the IT function.

Once you have developed your unique list of requirements, you can start looking for a suitable software solution that can deliver against those requirements and help make your strategic performance management initiative a success. In Fig. 10.16, I present a list of the leading vendors of performance management software applications (for the latest up-to-date list and links to their Web sites, please visit the resources section of the API Web site: www.ap-institute.com).

Finally, once you have shortlisted from among the possible vendors, always ask for reference clients that have implemented strategic performance management applications of similar scope and scale, and contact them. Many vendors will be happy to provide contact details, and a visit or conference call with other customers can be useful for both sides.

Vendor name	Product or solution name
360 Systems	Excelsis PMF
4GHI Solutions	Cockpit Communicator
Active Strategy	ActiveStrategy Enterprise
Actuate	Actuate Performance Management
AKS-Labs	Balanced Scorecard Designer
Applied PC systems	Strategy Map Balanced Scorecard Software
Clarity Systems	Clarity Performance Management Suite
Corda Technologies	CORDA CenterView, Corda PopChart
Corporater Group	Corporater Enterprise Performance Management Suite
CorVu	CorVu Performance Management Application
Covalent	Covalent Metrics, Projects, Risk, Core
Cubus	Corporate Overview, Strategic Project Performance Management
Escendency	Escendency System
Hitec (Laboratories)	Ten Performance Manager
HostAnalytics	Business Performance Management (BPM)
IBM	Corporate Performance Management, IBM Cognos Business Intelligence
i-nexus	Performance Manager, Program Manager, Process Manager
Infor	Infor PM (Performance Management)
InformationBuilders	Enterprise Business Intelligence
InPhase Software	Performance Plus
InsightFormation	InsightVision
Lawson	Enterprise Performance Management
Microsoft	Performance Point Server
MicroStrategy	Corporate Performance Management (CPM)
Nexala	Nexala Insights
Nexus	SPAR.net Performance Management
Oracle	Oracle Enterprise Manager, Hyperion Performance Scorecard
Procos	Strat&Go Performance Management
ProDacapo	Corporate Performance Management Suite
QPR Software	QPR Scorecard, QPR FactView, QPR Process Guide
Rocket Software	Gentia Enterprise Performance Management
SAP	SAP Strategy Management, Enterprise Performance Management
SAS Institute	SAS Strategic Performance Management
Stratsys	RunYourScorecard, RunYourCompany
Successfactors	SuccessFactor Performance Manager
Triangle	Performance Manager

FIGURE 10.16 List of leading vendors and software products.

SUMMARY

- The right performance management software applications are powerful enablers of performance management and bring the process to life. Specialized performance management software is an essential part of organizationwide attempts to measure and manage performance.
- At the same time, software applications are not a magic pill that can take away the hard work of designing unique and meaningful strategies and performance management information. A system will only be as good as the underlying approaches and information it supports. Remember, as with any software, the motto is: garbage in, garbage out!
- I recommended that you go for specialized performance management applications. The latest systems are so good and flexible that the immense costs and efforts of creating a home-grown system are not justified any more. Also, spreadsheet applications such as Microsoft Excel are not suitable applications for performance management.
- The key benefits of a performance management software application are (1) helping to engage all employees (and external stakeholders) in performance management through powerful communication and collaboration features, (2) providing all users with access to powerful performance management analytics and (3) creating a single integrated view of performance.
- I then provided advice on selecting the most suitable software application for your organization. I recommended clarifying the requirements and needs in terms of implementation scope, integration scope and scope for IT support.
- Finally, I outlined a framework and provided a template for defining your unique software requirements and supplied a list of the leading software vendors and solutions for measuring and managing performance.

REFERENCES AND ENDNOTES

1. See Marr, B. (2001). Scored for Life. *Financial Management*, 30(1), 14–17; Marr, B. and Neely, A. (2003). *Automating Your Scorecard: The Balanced Scorecard Software Report*. Gartner, Inc. and Cranfield School of Management, Stamford, CT; Marr, B. and Neely, A. (2001). *The Balanced Scorecard Software Report*. Gartner, Inc. and Cranfield School of Management, Stamford, CT.
2. Source: Marr, B. (2008). *Strategic Performance Management in Government and Public Sector Organizations*. The Advanced Performance Institute, UK; Marr, B. (2004). *Business Performance Management – The State of the Art*. Hyperion Solutions and Cranfield School of Management, London (both available at www.ap-institue.com).
3. See, for example: http://www.kpmg.co.za
4. de Waal, A. A. (2001). *Power of Performance Measurement: How Leading Companies Create Sustained Value*. Wiley, New York.
5. See also: Marr, B. and Neely, A. (2002). Software for Measuring Performance. In: *Handbook of Performance Measurement*, 2nd edn. (M. Bourne, ed.) pp. 210–41. Gee, London.

6. See, for example: Pendse, N. (2008). *The OLAP Report*. BARC (www.olapreport.com).

7. King, R., McIntyre, J., Moormann, M. and Walker, E. (1996). *A Formula of OLAP Success*. www.sas.com

8. This chapter is based on different market studies including: Marr, B. and Narr, D. (2004). *Software in Vergleich: 20 Werkzeuge for das Performance Management*. BARC, Germany; Marr, B. and Neely, A. (2003) (see note 1 above).

Chapter 11

Learning from Current Performance Management Practices

In this chapter, I take a look at the current state of practice in performance management. Everything I present here is based on the world's largest study of performance management in government and public sector organizations. The research project was recently conduced by the Advanced Performance Institute (API) in collaboration with the Chartered Institute of Public Finance and Accountancy (CIPFA) and sponsored by Actuate Corporation.[1] The research finds that organizations that apply the principles of strategic performance management outlined in this book outperform those who don't. The questions I will address in this final chapter are as follows:

- What do government and public sector organizations currently do in terms of performance management?
- Does performance management make a difference?
- What are the 10 principles of good performance management?
- Which of these principles have the strongest impact on performance?

A wide cross section of government and public sector organizations took part in this study including central or federal government agencies, state and local government bodies, as well as national health organizations, police forces, fire and rescue organizations, courts and education institutions. Overall, over 1100 respondents from the United States, Canada, Australia and the United Kingdom took part in this study. From each organization, we collected two responses, one from the chief executive (or equivalent) and the other from the person in charge of performance management.[2]

The findings of this comprehensive study confirm that just having a set of performance targets and performance measures in place does not lead to better performance. In fact, it often leads to a decrease in performance with perverse and dysfunctional behaviors such as suboptimization, target fixation, cheating and lying. However, applying the right principles and doing performance management properly do lead to better performance and more success. In the research, we identified 10 principles of performance management which make a real difference. Let me discuss each of these principles in further detail below.

THE 10 PRINCIPLES OF GOOD PERFORMANCE MANAGEMENT

The study identified 10 principles of good performance management for government and public sector organizations. Organizations that perform best are those which are able to (1) create clarity and agreement about the strategic aims, (2) collect meaningful and relevant performance indicators, (3) use these indicators to extract relevant insights, (4) create a positive culture of learning from performance information, (5) gain cross-organizational buy-in, (6) align other organizational activities with the strategic aims outlined in the performance management system, (7) keep the strategic objectives and performance indicators fresh and up-to-date, (8) report and communicate performance information well, (9) use the appropriate IT infrastructure to support their performance management activities and (10) give people enough time and resources to manage performance strategically(see Fig. 11.1).

Our research was able to validate that each of these 10 principles helps organizations to perform better. However, when they are all in place together, we can see the most significant impact on improved understanding, better decision making and superior performance. Therefore, addressing all these principles together far outweighs the impact of individual principles. The dark bars in the large arrow in Fig. 11.11 indicates the strength with which each of the principles impact performance; the thicker the bar the stronger the impact. The principles with the strongest individual impact on performance improvement are (1) creating clarity about the strategy with agreement on intended

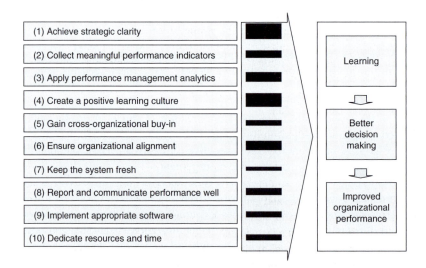

FIGURE 11.1 Ten principles of good performance management and their impact on performance.

outcomes, outputs and necessary enablers and (4) creating a positive culture of learning and improvement. Some of these impacts are indirect through improved learning, which leads to better decision making and which in turn leads to better performance. The big arrow indicates that the greatest impact is the combined impact of all of the principles together, which is far bigger than the sum of the individual impacts.

These findings are important in a world where the use of performance measurement and performance management has mushroomed among government and public sector organizations. The results of this study will hopefully help to confirm the principles outlined in this book and enable you to learn from it, put your own practices into context and improve your performance management approaches appropriately.

Achieving Strategic Clarity

I have discussed the principles of how to clarify your strategy in Chapters 1–4. In our research, we find that the majority of respondents (68%) agreed that articulating outcomes were more important than articulating outputs and that understanding and clarifying the key enablers were essential (94%) to improve performance. In fact, 89% of our respondents agreed that intangible elements were the most important enablers of future performance in their organization. We asked our respondents to rank the enablers in order of importance, and the five most important enablers identified were human capital, information technology, stakeholder relationships, data and information as well as corporate reputation and image.

We then explored how well the strategic objectives were identified. The majority of organizations articulate their output objectives while only just over half of them clearly articulate their outcome objectives. Less than half of the organizations clearly articulate enabling objectives and the most worrying finding is that 16% of organizations feel their objectives are not clearly articulated at all. This finding shows that even though respondents agree that outcomes and enablers are important, many fail to articulate objectives of these elements as part of their strategic plan.

> 16% of government and public sector organizations feel their objectives are not clearly articulated at all.

Throughout this book, I have discussed the importance of mapping strategies into cause-and-effect maps. Our survey finds that the majority of organizations (86%) in the government and public sector do not yet visualize their strategy using cause-and-effect maps. Ten percent of organizations create strategic cause-and-effect maps to visualize just the links between different perspectives, which is the most basic form of causal maps, and only 4% show the cause-and-effect linkages between their different strategic elements (see also Fig. 11.2).

There is also a lot of evidence that involving and engaging people in the strategy definition ensure that everybody is clear about the strategic direction.

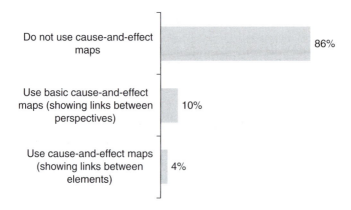

FIGURE 11.2   Cause-and-effect maps.

FIGURE 11.3   Engagement in strategy development.

In our research, we asked the respondents who in their organization was engaged in the formulation and definition of the strategy. Worryingly, in 6% of our organizations not even the chief executive was engaged in the strategy definition. My interpretation of this result is that maybe other government bodies specify performance goals and targets and the chief executive feels disengaged from the process. The other numbers are equally disappointing with only 89% of directors, 61% of senior managers, 21% of middle managers (a very low percentage) and a mere 2% of the rest of the employees feeling engaged in the process (see also Fig. 11.3). While it should be the role for senior executives and directors to formulate the strategy, this process should be one of engagement with the wider employees to ensure understanding and clarity about strategy.

Collecting Meaningful Performance Indicators

In Part II (Chapters 6–8) of this book, I have discussed importance of collecting the relevant performance information. Performance indicators should help you measure the things that matter the most and not those that don't. For that reason, performance indicators need to be linked to the strategic objectives of an organization. Only 15% of the respondents to our survey feel that all their indicators were linked to the strategy of their organization. And even less number of respondents (6%) believe that all their performance indicators are relevant and meaningful, whereas 92% feel that many of their indicators are not relevant and meaningful.

Part of the problem seems to be the large number of externally imposed indicators. Many organizations seem to make the assumption that these indicators are the only indicators they need to collect and that the set of externally dictated measures forms the core of their performance indicators. The problem with this is that many of the externally imposed measures and targets are output focused, and are collected for comparative purposes and to benchmark results with other government and public sector organizations. They often provide little insights about the unique strategic objectives of individual organizations and they don't help to measure the enablers. Our survey finds that the majority of organizations (39%) mainly measure outputs or a balance between outputs and enablers (29%). Only 23% believe they achieve a balance between measures for outcomes, outputs and enablers.

We asked organizations to name the broad perspectives they were using to measure their performance. We find that most organizations in the public and government sectors (98%) measure performance linked to people and human capital, these very often related to employee satisfaction, staff absenteeism and turnover as well as training. The second most popular dimension is resources-related measures such financial performance or infrastructure measures (76%). These are followed by regulator-related measures (68%), customer-related measures (only 57%), internal processes measures (52%), environmental and social measures (48%), and supplier-related measures (33%).

In addition to the fact that the majority of organizations believe their indicators are not linked to their strategy and are not very meaningful and relevant, 68% also feel they have too many indicators. Only 23% of organizations feel they have the right amount of performance indicators and 8% feel they have too few.

68% of government and public sector organizations feel they have too many indicators.

Well-designed indicators should help organizations to extract meaningful insights and management information that help them to learn and improve performance. For that reason, indicators should have clear, realistic and achievable (though stretching) targets. In our survey, we discover that the majority of organizations feel that their target-setting process is unclear (82%) and many believe that set targets are unrealistic and unachievable (53%).

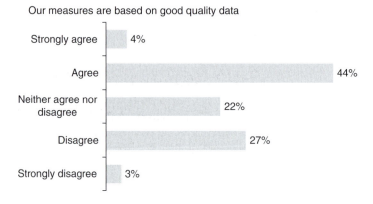

FIGURE 11.4 Data quality.

Another problem seems to be the mechanistic application of measurement with a focus on collecting data on everything that is easy to count. Few reflections seem to take place on the best ways of measuring performance to achieve the relevant and meaningful insights. This is partly driven by the large number of mandatory indicators that leave little or no room for experimentation with and improvement of measurement techniques. However, if organizations want to get to grips with performance and especially with their intangible enablers, then they need to allow experimentation with measurement and encourage innovative ways of data collection. Only 8% of organizations feel they give people the freedom to change or challenge the measurement methods, and most people feel disengaged from the measurement design and target-setting process.

The other problem is that organizations might measure the 'right' things but can't get the data to adequately assess it. Data quality is still an issue in many public sector and government organizations, especially when it comes to some of the nonfinancial indicators where we often rely on simple surrogate measures. About a third of respondents feel their data quality is inadequate and another 22% are unsure about the data quality (see Fig. 11.4).

Applying Performance Management Analytics

Once organizations have collected meaningful data, they can analyze it and gain insights about their performance. Performance management analytics are tools and techniques that allow organizations to convert their performance data into relevant information and knowledge. Without it, the entire performance management exercise is of little or no value to the organization. We see that many organizations seem to spend the majority of their time and effort on collecting and reporting data and don't give enough time on extracting valuable and actionable insights from their performance data. An important finding of this research is that many respondents (59%) feel that their organization does

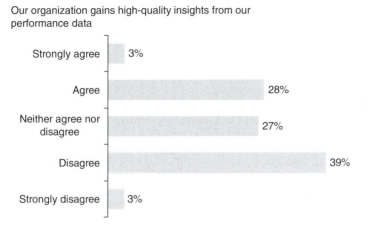

FIGURE 11.5   Insights from data.

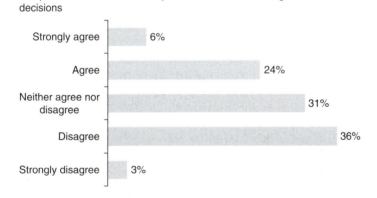

FIGURE 11.6   Strategic decision-making.

not have sufficient capabilities to comprehensively analyze performance data and 87% feel that their analysis capabilities need to be improved.

Our survey finds that the majority of public sector and government organizations are unhappy with the insights they gain from their performance data. Only about a third of organizations in our survey report that they gain high-quality insights from their performance data, while 42% are unhappy with the insights that are extracted from the performance data (see Fig. 11.5).

Good-quality insights from data should support better-informed decision making. We asked our participants whether their performance indicators help them to make better strategic and operational decisions (see Figs 11.6 and 11.7). Overall, the picture is not satisfactory as the majority of organizations

FIGURE 11.7   Operational decision-making.

feel that their performance data does not support their decision making. The findings show that the performance measures support operational decision making slightly better than strategic decision making, which indicates that either the data is too operational or that the analysis is not taken to a strategic level.

Benchmarking can be used to analyze performance and learn by comparing own performance levels to those of others. Benchmarking can be done internally to compare departments as well as externally to compare different organizations. The majority of the organizations in this study (51%) use external benchmarking, while 26% use internal benchmarking and 23% do no benchmarking. These results seem surprising especially because so much comparative data is publicly available.

The strategy of an organization with its outcomes, outputs and enablers is basically a set of assumptions about cause-and-effect relationships. Performance indicators can ultimately be used to test and validate these assumptions and relationships. The indicator data has been applied in commercial organizations to validate relationships between, for example, staff satisfaction, client satisfaction and profits. However, our data shows that only 13% of public sector and government organizations claim to use their performance data to test causal relationships.

Creating a Positive Learning Culture

In Chapter 9, I have outlined the importance of a performance-driven culture. Performance information can be used for different purposes. It can be used to blame or punish people or it can be used to empower people and enable self-management and learning. The initial signs were positive in our survey. Eighty-three percent of organizations stated that they place a high priority on learning from performance information. However, some underpinning data calls this figure into question. A strong indicator for a failing performance management culture is the fact that data is fabricated or made up to fulfill reporting and target requirements. We discover that 68% of public sector organizations occasionally make up or fabricate performance data. This is a staggeringly high number, which clearly points away from a positive learning culture.

68% of public sector organizations occasionally make up or fabricate performance data.

To investigate this further, we asked organizations about the way directors and senior managers are using performance measurement information.[3] The majority of organizations feel that their directors and senior managers guide employee's behavior by using performance indicators to monitor the implementation of objectives (providing feedback on the achievement of set goals to control and correct unwanted variance). This is called diagnostic control, which many now see as inappropriate for modern organizations as it can cause all the negative consequences discussed in Chapters 6 and 8. The second most frequently used approach is to focus on prescribing overall strategic purpose, vision and values. This is less directive and instead focuses on creating an understanding of strategic intent. The least frequently used approach was one in which directors and senior managers involved themselves regularly and personally in the decision making activities (drawing attention toward current strategic issues). Current research favors a mix of the two latter control levers, where top management sets the direction and then influences direction through personal interaction with people.

Another factor that has an impact on how successful performance management can be implemented is whether top management focuses on controlling outputs, inputs or behaviors. Again, research has shown that controlling people's behavior in a top-down manner where the emphasis is on compliance with articulated operating procedures yields the least benefits. The problems here are the huge costs involved in surveillance and control, the fact that it reduces discretion and leads to rigid and often dysfunctional behavior.[4] Focusing on controlling outputs provides subordinates discretion of how to achieve the desired ends. Output control makes sense in the absence of cause-and-effect knowledge; however, it can lead to an overemphasis on short-term targets such as financial performance to the detriment of longer-term objectives. Input control on the other hand focuses on the antecedents of performance. For this control focus causal links need to be known, otherwise there is a danger that the wrong performance drivers might be managed. In terms of best practice, public sector and government organizations should have a clear understanding of cause-and-effect and then use a mix of input and output controls to mange their organizations. In our sample, the majority of respondents feel that their directors and senior managers focus on outputs, followed by a focus on behavioral control, and a focus on inputs.

In a positive learning culture, performance information and contextualized feedback are provided to everyone in the organization, with a special emphasis on middle management and frontline staff. Other companies mainly provide information to senior management and external stakeholders. Unfortunately, this is what the majority of organizations in our sample focus on. This further supports for the fact that the main focus for most organizations is not on organization-wide learning. Organizations that want to provide everyone in the organization with relevant performance information would focus on contextualized feedback on how the organization is performing and how this can be improved. Overall, the above analysis shows that many public sector organizations have quite a long way to go until they create the right environment for good performance management. It is evident that there is currently a too strong bias on command and control and not enough performance feedback to middle managers and frontline staff.

Gaining Cross-Organizational Buy-in

In Chapter 9, I have discussed the importance of performance management buy-in. To make performance management work, buy-in is required across the organization. Top-level buy-in is essential to get the system designed and implemented. Buy-in from middle managers and frontline staff is essential to make performance management an internal part of the organization's daily routines and ensure ongoing use and value.

The organizational culture and approach toward performance management tend to also have a strong impact on buy-in. If middle managers and frontline staff are not provided with meaningful performance feedback and instead feel treated like robots that just have to perform prescribed tasks or collect and report seemingly meaningless data, then they are unlikely to ever buy-in to the performance management approach.

In our survey, we asked each respondent to identify the level of awareness, use and acceptance of the performance management system among the different employee groups. What we find is that the level of acceptance, use and awareness goes down the lower you go down the organizational hierarchy. This is in line with what other studies have found. The highest levels of acceptance and use are among the chief executives and directors (see Fig. 11.8). What is surprising and worrying is the low level of usage among middle managers and frontline staff, of which a considerable number aren't even aware of the performance management system.

This is an important finding and calls for government and public sector organizations to find better ways of engaging their middle managers and frontline staff in their performance management.

Ensuring Organizational Alignment

In Chapter 5, I outlined that not linking and aligning performance management with other management processes in an organization can severely reduce its

Our different employee groups have the following levels of awareness, use and acceptance of our performance management system.

	Clear advocate of the PM system	Accepts the PM system, but isn't an advocate	Uses the PM system, but doesn't fully accept it	Aware of the PM system, but doesn't use it	Not aware of the PM system
Chief executive	55%	36%	6%	3%	0%
Directors and executive managers	33%	48%	14%	5%	0%
Senior managers	26%	40%	23%	8%	3%
Middle managers	4%	38%	32%	20%	6%
Rest of employees	2%	22%	15%	30%	31%

FIGURE 11.8 Awareness, use and acceptance of performance management.

benefits. It makes sense to use the strategic performance management system to guide and align other organizational processes such as budgeting, performance reporting, management of projects and programs and management of risks.

In many government and public sector organizations, reporting, risk management, project management and budgeting processes are run in parallel to their strategic performance management approaches. Our research finds that alignment is still a major problem in public sector and government organizations. Seventy-three percent organizations have aligned their performance reporting, while less than half have aligned their budgeting and project and program management. Only 14% of organizations believe their risk management is fully aligned with their performance management system (see Fig. 11.9).

Keeping the Performance Management System Fresh

The strategy of an organization has a shelf life and has to be revised and amended to ensure it stays relevant. I have discussed this in Part I of this book and as part of the strategy revision meetings in Chapter 9. In the same way, the performance management system has to be revised and kept fresh. If the strategic objectives change, the performance indicators should change. However, I often see organizations that build up huge legacy systems of performance indicators because they keep adding new ones but never delete the obsolete ones.

Our performance management system is fully aligned and integrated with:

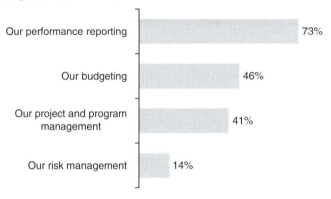

FIGURE 11.9   Performance management alignment.

To stay relevant, the strategic objectives (outcomes, outputs and enablers) have to be reviewed regularly and the performance indicators have to be revised accordingly.

The findings from our survey indicate that government and public sector organizations regularly review and renew their strategy. Most public sector organizations follow an annual planning cycle for which they need to create a business plan and therefore have to review and renew their strategy. In a similar way, we find that respondents feel that their performance management system and their performance indicators are regularly reviewed and renewed.

Reporting and Communicating Performance Information Appropriately

In Chapters 9 and 10, I have outlined the importance of reporting and communicating performance information appropriately. It provides people with the insights they require to inform their decision making and learning. In our study, we find that the primary communication format is numeric, using tables and spreadsheets complemented by graphs and charts. This is followed by pure numeric without the graphs and charts. The least common formats were narratives with supporting numeric data and verbal communications of performance information. I believe that organizations should place much more emphasis on communicating performance information in words, both written and verbal, and less in numbers. The underlying messages and insights the numbers generate are what really count. The respondents to our survey seem to agree with this because the majority is unhappy with their current communication format. Over a third feel that their current communication format is not appropriate or meaningful (see Fig. 11.10).

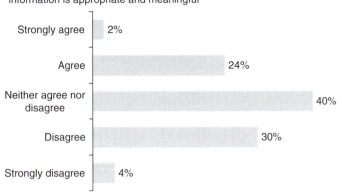

The communication format of our current performance information is appropriate and meaningful

FIGURE 11.10   Communication of performance information.

Implementing Appropriate Software

In Chapter 10, I have outlined the benefits of leveraging specialized performance management software applications to bring the process to life. In our survey, we find that over half of the organizations still rely on spreadsheet applications, while 23% use specialized packaged applications and 22% use custom-build applications.

We also explored how well the different software applications support performance management. Figure 11.11 indicates that users of packaged applications are significantly happier with the way their software supports performance management in their organization, followed by custom-build applications and spreadsheets.

We then explored how well the software application helps to engage everyone in the process of managing and measuring performance. Figure 11.12 indicates that users of packaged applications are again significantly happier with the way their software helps to engage people in the performance management activities, followed by custom-build applications and spreadsheets, as expected.

Dedicating Resources and Time to Performance Management

Finally, to make performance management work, organizations have to dedicate resources and time to the process. Good performance management will not just happen. Processes need to be embedded into the organization; measures need to be collected, analyzed and reported; people have to be trained; meetings have to take place to discuss performance; the IT system has to be maintained; and the system has to be kept fresh and updated. All of this

The software application(s) we use for performance management
helps us to better manage our performance

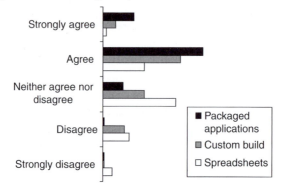

FIGURE 11.11   Software application and performance management.

The software application(s) we use for performance management
helps us to better engage everyone in the process of managing and
measuring performance

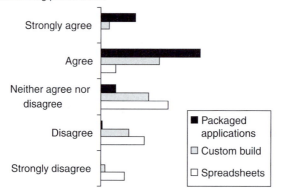

FIGURE 11.12   Software application and engagement in performance management.

requires time and resources, which both tend to be on short supply in any
organization.

Current best practice is to create a dedicated team with resources and time
to perform the role of facilitating the performance management activities. This
team then performs the tasks such as facilitating the strategy design and mapping process, designing and reviewing performance indicators, collecting and
analyzing performance data, reporting performance, facilitating the cascade and
the performance review processes, maintaining the performance management
software system as well as training people in the performance management

process. This is a broad remit and requires a cross-functional team to make it work. This team should also not just be a project team that only exists for a few months or so but should be established for good. While the focus of the activities might change over the maturity of the performance management system, there will always be the need for dedicated people to make performance management work.

Interestingly, most of the government and public sector organizations in our study have dedicated people in place to do it. Job titles such as director of performance, performance officer and performance analyst are nowadays commonplace in the public sector. Even though these posts are in place, there is still a recognized need for more resources, capabilities and a wider remit. Eighty-two percent of organizations had a dedicated person (part or full time) who was in charge of performance management and 73% stated that their remits were too narrow, 87% feel that their analysis capabilities need to be improved and 85% feel they require more resources and a larger team.

SUMMARY

- In this chapter, I have presented the findings from the world's largest study of performance management practice in government and public sector organizations.
- The research confirms that performance management, as long as it is done well, has a significant impact on performance. Organizations that apply the principles of good performance management practices outperform their peers.
- The study identified the principles of good performance management, which are all supported by the arguments in this book.
- Achieving strategic clarity and creating a positive learning culture are the two most important success factors. However, all 10 together have the greatest impact.
- Overall, it shows that there is a lot of work to do in most government and public sector organizations.
- The advice, tools and templates in this book have been designed to close the gaps identified in practice and should enable government, public sector and not-for-profit organizations to measure and manage what matters.

One final note, please get in touch if you have any comments or suggestions about the material I have introduced in this book. I am keen to hear whether the tools have made a difference or if anything hasn't worked in some circumstances. Also, please let me know if you have suggestions about how to make any of the tools or template better and contact me if you require more information or if you feel I – or one of my colleagues at the Advanced Performance Institute – could help with any of your strategic performance management challenges. You can reach me at: bernard.marr@ap-institute.com or you can contact us via the API Web site: www.ap-institue.com.

REFERENCES AND ENDNOTES

1. Marr, B. (2008). *Strategic Performance Management in Government and Public Sector Organizations*. Research Report, the Advanced Performance Institute, UK. The full report can be downloaded for free at: www.ap-insitute.com

2. For a detailed break down of respondents please refer to the research paper: Marr, B. (2008). *Strategic Performance Management in Government and Public Sector Organizations*. Research Report, the Advanced Performance Institute, UK.

3. These items are based on the research of: Simons, R. (1995). Control in an Age of Empowerment. *Harvard Business Review*, 73(2), 80–88.

4. See for example: Snell, S. (1992). Control Theory in Strategic Human Resource Management: The Mediating Effects of Administrative Information. *Academy of Management Journal*, 35(2), 202–325; Hofstede, G. (1978). The Poverty of Management Control Philosophy. *Academy of Management Review*, 3(3), 450–461.

Go to www.ap-institute.com to download the latest:

- articles,
- management white papers,
- case studies,
- research reports.

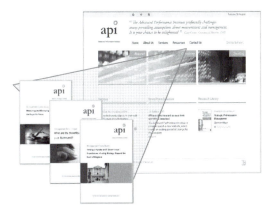

OTHER RECENT BOOKS BY BERNARD MARR

Strategic Performance Management

This book is about strategic performance management for the twenty-first century company. This practical book is more suitable for commercial managers. 'Managing and Delivering Performance' is the public sector and not-for-profit version of this book which navigates readers though the identification, measurement and management of the strategic value drivers as enables of superior performance. Using many real-life case examples, this book outlines how organizations can visualize their value creation, design relevant and meaningful performance indicators to assess performance and then use them to extract real management insights and improve everyday strategic decision making as well as organizational learning.

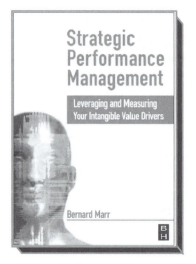

'Strategic Performance Management'
Published by Butterworth-Heinemann (2006)
ISBN-13: 978-0750663922

Perspectives on Intellectual Capital

This book is about managing, measuring and reporting of intangible assets and intellectual capital. Intangibles are the key drivers of performance in most modern organizations, and many public sector organizations now realize that they need to manage, measure and report on those elusive enablers of future success. This book includes contributions from the leading thinkers in the field of intellectual capital to provide a cross-disciplinary overview of this increasingly important topic.

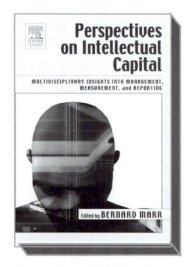

'Perspectives on Intellectual Capital'
Published by Elsevier (2005)
ISBN: 0-7506-7799-6

Index

A

Aberdeen Group, 223
Agency theory, 142
Analysis functionality, of performance management software applications, 265
Argyris, Chris, 214
Assessing partnership performance, 165
Atkinson, Philip, 211
Austin, Robert, 145
Automation choices, performance management, 260–263
  customization, 262
  implementation, 261
  integration, 261–262
  IT support, 262

B

Balanced scorecard model (BSC)
  building of VCM and, 72
  perspectives, 69
  strategy map template of, 71
  traditional, 70
  versus value creation maps, 76–78
Belfast City Council, VCM/VCN practices, 87–90
BI. *see* Business intelligence (BI)
Biased command-and-control measurement systems, 144
Black, Paul, 217
Bottleneck risk, 118
Boyle, David, 138
Bruijn, Hans de, 140, 153
BSC. *see* Balanced scorecard model (BSC)
Budgeting process
  aligning, 110–112
  case study, 112
Bundled value creation map, 74–75
Business activities
  aligning, 107–110
  mapping existing and planned activities, 108–109
  prioritizing, 110
Business intelligence (BI), 256
Business planning
  aligning activities, 107–110

C

Causal value creation map, 75
Cause-and-effect relationships, 74–75
Cheating
  performance indicator, 190
  performance indicator decision framework, 182
Cheat proof measurement system, 145
Citywide performance reporting (CPR), 146
Collaboration, lack of, in spreadsheets, 247
Color-coded map, 257
Commentary, organizational performance, 225–226
Communication, effective
  best practices, organizational performance, 224–228
  commentary, 225–226
  provide numerical data, 226
  use good communication channels, 226–228
  visualize data in graphs, 226
Communication support, lack of, in spreadsheets, 247
Competition risks, 116–117
Compliance, 137
  measuring for, 145–150
COMPSTAT meetings, 234
Confidence level, performance indicator, 189
Controlling behavior, 137, 142–145
CPR. *see* Citywide performance reporting (CPR)
Cross-departmental structure, organizational structure, 113

Cross-functional structure, organizational structure, 113

Customization, performance management software applications, 262, 266

Customized report, 253

D

Dashboards, 252, 253

Data collection, 163–164

Data collection, performance indicator design template, 185–187
date and frequency of, 186–187
formula/scale/assessment, 185–186
Likert scale, 186
measurement instrument, 185
triangulation, 194–195

Decision making, performance indicator decision framework, 181–182

Difficult analysis approach, of spreadsheets, 247

Disaster risk, 118

Double-loop learning, organization, 215

Dysfunctional behavior, 143
performance indicator, 190

E

Ehin, Charles, 145

Emphasizing learning, performance-driven culture, 216–218
formative assessment, 217
summative assessments, 217, 218

Employee theft risk, 119–120

Empowerment, 137–138
measurement for, 150–151

External reporting, 137
measuring for, 145–150

External risks
competition risks, 116–117
market risks, 117

Extrinsic rewards, 222–223

F

Financial risks, 117

Finkelstein, Sydney, 167

Focus groups, performance assessment, 193

Ford motor company, 166

Formative assessment, emphasizing learning, 217

G

Gardner's studies, 139

Good targets, organization, 188

Google, 162

Government organizations monetary rewards in, 223

Graphical user interface (GUI), 256

GUI. *see* Graphical user interface (GUI)

H

Handy, Charles, 154

Heat map, 109

Home-grown, performance management software applications, 260

Home Office, VCM/VCN practices, 93–94

Hook, Dee, 154

Hubbard, Douglas, 176, 178

Human resources, risks to
employee theft risk, 119–120
knowledge risk, 119
staffing risk, 119

I

IBN, 192, 211, 221

Impact analysis, 258

Implementation, performance management software applications, 260, 261

Index, for performance indicator, 195

Influence, 74

Informational measurement, 145

Insurance Mutual (TT Club), VCM/VCN practices, 96–98

Integration, performance management software applications, 260, 261–262

Intellectual property rights theft risk, 120–121

Intelligence quotient (IQ), 139

Interpretation of PI, 166

Interviews, for performance indicator, 192–193

Intrinsic rewards, 222–223

IQ. *see* Intelligence quotient (IQ)

IQ test, 139, 152

IT support, performance management software applications, 262

IT systems risk, 120

K

Kaplan, Robert S., 69, 215
Kelly, D. Lynn, 152
Key performance indicators (KPI), 165
Key performance questions (KPQ), 151,
    176
  creation, 167–172
    challenge existing PI, 171–172
    for communicate and review
      performance, 172
    design performance indicator, 171
    engage people in, 169–170
    focused on present and future, 171
    open questions, 170–171
    for performance related, 169
    short and clear, 170
    strategic objective on VCM, 168
  guides us to right indicators, 164–165
  help us to interpret, 166
  help us to learn, 162
  more important than KPIs, 173
  and performance indicator decision
    framework, 179
  performance indicator design template,
    185
  power of, 166–167
  practical examples, 172–173
  systematic approach to data collection,
    163–164
Knowledge risk, 119
KPI. see Key performance indicators (KPI)
KPQ. see Key performance questions
    (KPQ)

L

Leadership, performance-driven,
    221–222
Learning. see Organizational learning
Likert scale, performance indicator design
    template, 186

M

Mapping
  existing and planned activities, 108–109
Market risks, 117
Marquardt, Michael, 167
Matrix structure, organizational structure,
    113

Measurement instrument
  decision on, 191–194
  performance indicator design template,
    185
Measurement systems, performance
  in call centres, 143
  cheat proof, 145
Meetings, organizational, 228–237
  characteristics of, performance-driven
    culture, 235–237
  operational performance improvement,
    233–234
  personal performance improvement,
    234–235
  strategic performance improvement, 233
  strategy revision, 231–233
Memory losing risk, of spreadsheets, 247
Meyer, Marshall, 143
Microsoft Excel, 246
Motor Neurone Disease Association,
    VCM/VCN practices, 98–99
Multidimensional analysis, 259
  tools, 256
Mystery shopping approaches, for
    performance assessment, 193

N

National Lottery Commission, VCM/
    VCN practices, 100–101
Norton, David P., 69, 215
Not-for-profit organizations
  monetary rewards in, 223
Notification, performance indicator, 191
Numerical data, for organizational
    performance, 226

O

Open vs closed questions, 170
Operational performance improvement
    meeting, 233–234
  effective crime-fighting strategy in,
    234
Organization
  performance-driven culture of. see
    Performance-driven culture
  performance indicator. see Performance
    indicator
  target-setting process of, 187–189

Organizational learning, 137–138
  double-loop learning, 215
  measurement for, 150–151
  and performance-driven culture,
    213–215
  performance management, 214–215
  single-loop learning, 214
  triple-loop learning, 215
Organizational performance
  out comes of, 145
  reporting, best practices, 224–228
    commentary, 225–226
    provide numerical data, 226
    use good communication channels,
      226–228
    visualize data in graphs, 226
  scenarios of meetings for, 229–230
Organizational structure
  aligning, 112–113
  types of, 113
Ornstein, Robert, 68

P

Partnership-based delivery model, 163
Peer-to-peer evaluation, for performance
  assessment, 194
Performance assessment, 151, 176
  assumptions, 176
  case study, 195–205
  defined, 152–153
  dos and don'ts, 153–155
  external assessments, 193
  focus groups in, 193
  in-depth interviews for, 192–193
  index creation for, 195
  mystery shopping approaches for, 193
  observations, 193–194
  peer-to-peer evaluation, 194
  surveys and questionnaires for, 192
Performance data analysis, 256–259
Performance-driven culture
  building blocks for, 212–213, 220
  characteristics of performance
    meetings, 235–237
  creation of, 216
  defined, 212–213
  emphasizing learning and performance
    improvement, 216–218

formative assessment, 217
  summative assessments, 217, 218
  four enablers of, 218–220
  key behaviors of senior leaders, 221
  organizational learning and, 213–215
  review and discuss performance
    interactively, 228–231
  reward and recognize people, 222–224
  tips for, 223–224
Performance-driven leadership, 221–222
Performance improvement
  meetings. *see* Performance
    improvement meetings
  performance-driven culture, 216–218
Performance improvement meetings
  characteristics in performance-driven
    culture, 235–237
  operational, 233–234
    effective crime-fighting strategy, 234
  personal, 234–235
  strategic, 233
Performance indicator decision
    framework, 178–183
  cheating, 182
  collecting indicator, 183
  collecting meaningful data, 178–179
  decision making, 181–182
  decisions of potential indicator,
    180–181
  indicators assessment cost, 182
  KPQ and, 179, 181
  potential indicator, 180–181
    decisions of, 180
    use of existing methods for,
      180–181
  strategic element in, 178–179
Performance indicator design template,
    183–184
  five elements of, 184–185
  KPQ and, 185
  methods for data collection, 185–187
Performance indicators, 137, 151
  assessment cost, 182
  assumptions, 176
  collecting, 176
  confidence level, 189
  cost estimation in, 189
  decision framework, 178–183

defined, 152, 175–176
design template. *see* Performance
    indicator design template
indirect and proxy, 177–178
possible dysfunctions by, 190
reporting, 190–191
target-setting process in organizations,
    187–189
triangulate, 194–195
Performance management
    key behaviors of senior leaders and, 221
    organizational learning, 214–215
Performance management software
    applications, 245–267
    advantages, 247–259
        engaging people, in performance
            management, 249–255
        performance data analysis,
            256–259
        single integrated view creation, of
            performance, 248–249
    selection, 259–267
        automation choices, 260–263
        buying, 260
        framework, 263–267
Performance measurement
    for controlling behaviour, 143–145
    defined, 151–153
    implications for, 141–142
    on public sectors, 140
    reasons for, 137–138
    systems, 142
Performance-related vs strategic-choice
    questions, 169
Personalized homepage dashboard, 253
Personal performance improvement
    meeting, 234–235
Peterson, Donald, 166
Physical resources, risks to
    bottleneck risk, 118
    disaster risk, 118
Police dashboard, 252
Poor targets, organization, 189
Pricing, performance management
    software applications, 263–265
Project management module, 255
Public sectors, performance
    measurement, 140

R
Recognition, performance, 222–224
Regulatory risk, 121
Relational resources, risks to
    reputation risk, 122
    supply chain risk, 122–123
Reporting
    best practices, organizational
        performance, 224–228
        commentary, 225–226
        provide numerical data, 226
        use good communication channels,
            226–228
        visualize data in graphs, 226
    performance indicator, 190–191
Reputation risk, 122
Reward, performance, 222–224
Risher, Howard, 212
Risk management
    aligning, 113–115
    case study, 114–115
Risk management module, 255
Risks
    analysis, 116–125
    external, 116–117
    financial, 117
    human resources. *see* Human
        resources, risks to
    management. *see* Risk management
    physical resources. *see* Physical
        resources, risks to
    relational resources. *see* Relational
        resources, risks to
    structural resources. *see* Structural
        resources, risks to
Root-cause analysis, of performance,
    256
Royal Air Force, VCM/VCN practices,
    90–92

S
Sagan, Carl, 165
Scalability, lack of, in spreadsheets,
    247
Schmidt, Eric, 162
Scientific method, 163
Scottish Intellectual Assets Centre, VCM/
    VCN practices, 94–96

Selection framework, for performance
management software
applications, 263–267
analysis functionality, 265
collaboration, 266
communication, 266
flexibility, 266
future vision, 267
pricing, 263–265
scalability, 265
security, 266
services, 267
technical specifications, 266–267
user interface/data presentation,
265
vendor information, 265
Single integrated view creation, of
performance, 248–249
Single-loop learning, organization, 214
Software applications, for performance
management, 245–267
Software-selection matrix, performance
management, 264
Sperry, Roger, 68
Spitzer, Dean, 192, 211, 221, 228
Spreadsheet applications, disadvantages,
246–247
Staffing risk, 119
Strategic element
performance indicator decision
framework, 178–179
performance indicator design template,
184–185
assessment, 184
ownership, 185
Strategic performance improvement
meeting, 233
Strategic performance management, and
software applications, 246
Strategy
key elements of organizational, 72–73
value creation map (VCM), 68–69
value creation narrative (VCN), 68–69
Strategy map template
Kaplan and Norton's, 71
public sector and not-for-profit, 71
Strategy revision meeting, 231–233
Structural resources, risks to

intellectual property rights theft risk,
120–121
IT systems risk, 120
regulatory risk, 121
Suboptimal performance, 143
Summative assessments, emphasizing
learning, 217, 218
Supply chain risk, 122–123
Surveys and questionnaires, for
performance assessment, 192
Systematic approach to data collection,
163–164

T
Target-setting process, organization,
187–189
good/poor targets, 188–189
targets and performance thresholds,
188–189
tips for, 188
Taylor, Frederick, 142
Technical specifications, of performance
management software
applications, 266–267
Thurber, James, 173
Time-consuming approach, of
spreadsheets, 247
Top-down control measures, 153
Triangulate indicators, 194–195
Triangulation, performance data
collection, 194–195
Triple-loop learning, organization, 215

U
Usage of measures, in organizations,
141–142
User interface/data presentation, for
performance management
software applications, 265

V
Value creation maps
color-coded fields, with, 250
hierarchical tree diagram, with, 251
speedometer displays, with, 250
Value creation map (VCM), 161
versus balanced scorecard, 76–78
basic template of, 74

BSC and, 72
bundled, 74–75
cascading of, 84–87
case study, 87–101
causal, 75
cause-and-effect relationships, 74–75
creation of, 78–83
mapping existing and planned activities
      onto, 108–109
organizational practice, 87
performance levels, 109
strategic elements and, 72–73
Value creation narrative (VCN)
      case study, 87–101
      creation of, 84
      definition, 83
      organizational practice, 87

VCM. *see* Value creation map (VCM)
VCN. *see* Value creation narrative (VCN)
Vendor information, for performance
      management software
      applications, 265, 268
Visa network, 154
Visual data, for organizational
      performance, 226

W
Waal, André de, 248
William, Dylan, 217
Workflow, performance indicator, 191

Y
Yankelovich, Daniel, 140